The Handboo
Market Research for Life Science Companies

There are times innovation emerges from a lab before the market has even been defined. When that happens, the market must be identified and mapped through iterative steps.

This book is a practical playbook for researchers and founders navigating that process, from defining your first market hypothesis to estimating your market size, to collecting and analyzing your own data to find product-market fit.

Building on the award-winning first edition, this second edition adds new frameworks for using AI responsibly in research, managing bias in data analysis, harnessing the importance of storytelling and building the continuous research mindset that emerging technologies demand.

The Handbook of Market Research for Life Science Companies

Navigating the Gap Between Lab and Market

Second Edition

Jean-François Denault

Routledge
Taylor & Francis Group
A PRODUCTIVITY PRESS BOOK

Designed cover image: Shutterstock

Second edition published 2027
by Routledge
605 Third Avenue, New York, NY 10158

and by Routledge
4 Park Square, Milton Park, Abingdon, Oxon, OX14 4RN

First edition published 2018 by Routledge

Routledge is an imprint of the Taylor & Francis Group, an informa business

Library of Congress Cataloging-in-Publication Data
Names: Denault, Jean-Francois author
Title: The handbook of market research for life science companies : navigating the gap between lab and market / Jean-Francois Denault.
Description: Second edition. | New York, NY : Routledge, 2027. | Includes bibliographical references and index.
Identifiers: LCCN 2026014408 (print) | LCCN 2026014409 (ebook) | ISBN 9781041223245 hardback | ISBN 9781041223221 paperback | ISBN 9781003731726 ebook
Subjects: LCSH: Biotechnology industries—Marketing | Life sciences—Economic aspects
Classification: LCC HD9999.B442 .D464 2027 (print) | LCC HD9999.B442 (ebook) | DDC 660.6068/8—dc23/eng/20260529
LC record available at https://lccn.loc.gov/2026014408
LC ebook record available at https://lccn.loc.gov/2026014409

ISBN: 978-1-041-22324-5 (hbk)
ISBN: 978-1-041-22322-1 (pbk)
ISBN: 978-1-003-73172-6 (ebk)

DOI: 10.4324/ 9781003731726

Typeset in Garamond
by Apex CoVantage, LLC

Contents

Preface

Market research is a well-documented business activity. If you are looking for information on market research, you will find many well-written resources and books on the topic.

So, why write a book specifically for startup companies in life sciences?

Most of the time, life science startups are trailblazing. They develop new technologies with no defined market. They redefine markets by developing a technology that goes across multiple verticals. As these startups perpetually redefine science and technology, they constantly reshape and redefine the markets in which they exist.

As such, a comprehensive understanding of the market is crucial. Investors and partners seek companies with an accurate read of their market, and pitching to them with a great technology but with a poor understanding of the market, and a poor grasp on how this technology serves the market is misguided.

This book introduces market research for those who need a practical playbook and who are active in an emerging market. It is intended for:

- *Entrepreneurial academics, startups, or small companies:* Looking to define their markets and needing tools to define their market
- *Established market research practitioners:* Looking to increase their knowledge of market research for the life sciences industry

I want to thank my clients, with whom I have had the opportunity to work for many years. For some, our relationship goes back 20 years; for others, our journey is just beginning. From all of you, I have learned so much and had the opportunity to work on so many different projects; this book would not have been possible without these opportunities.

I also want to take a moment to thank Ajit for his time and effort on this edition. In the first edition, he graciously agreed to be interviewed and help shape the content. For this one, he went above and beyond by not only crafting a powerfully written foreword but also taking the time to read and review the first draft of this work. His feedback and advice continue to be appreciated.

I also want to take a moment to thank Emily for her insights on delivering difficult truths, an invaluable skill in market research.

Finally, profound thanks to my wife, Corinne, for her help with revisions and suggestions, and to my kids, Jaz and Chad, and my parents, Daisy and Jean-Marc, for their support throughout this project.

It is my hope that this book can help you on your journey to innovation.

About the Author

Jean-François Denault is the author of three books on life sciences commercialization and market research, published by CRC Press and Productivity Press (Taylor & Francis). His previous work, *Innovate, Fund, Thrive: The Entrepreneur's Playbook to VC Fundraising in Life Sciences* (2024, co-authored with Philippe Tramoy), builds on his previous books: *The Handbook of Marketing Strategy for Life Science Companies* (2018) and *The Handbook of Market Research for Life Science Companies* (2017), which won the American Library Association's Outstanding Business Reference Sources Award in 2018.

With 25 years of consulting experience, Jean-François has completed more than 450 projects across 32 countries for Fortune 500 companies—including Johnson & Johnson, Novo Nordisk, AbbVie, Procter & Gamble, Microsoft, and Adobe—as well as government agencies and over 100 innovative life sciences startups. He specializes in market research, competitive intelligence, and commercialization strategy for pharmaceutical, biotechnology, and medical device companies.

In recent years, he has worked with incubators and accelerators across Canada. Jean-François serves as Expert in Residence at the Quebec Biotechnology Incubator (CQIB) and mentored over 75 startups through leading programs including Next AI, HEC Montréal, the European Institute of Innovation and Technology, and Connect'Innov (Tunisia). He sits on the editorial boards of the *Journal of Brand Strategy* and *Applied Marketing Analytics.*

He holds an Executive MBA in Bio-Industry from UQAM and graduate degrees in Management Consulting and Organizational Communication, and undergraduate degrees in Finance, International Commerce, and East Asian Studies (Chinese concentration). Based in Montreal, Canada, he can be reached at jf.denault@impacts.ca or through LinkedIn at linkedin.com/in/jfdenault.

Foreword

In 1984, my friend Dr. Mike Phelps, inventor of positron emission tomography (PET) and the co-founder of CTI (a company that was later acquired by Siemens), was told by his market research team that there was a market for fewer than 50 PET systems in the USA. Mike's response was "*you are off by at least one order of magnitude.*" The proverbial tension between intuition and analysis followed and they agreed to disagree.

As of the last count, there are 1,500+ PET or PET-CT systems installed in the USA.

Market research has always occupied an ambiguous position in life sciences. It is indispensable, yet frequently misunderstood. On one extreme, it is a disciplined way of asking questions about the world's needs, constraints, trade-offs, and possibilities. At the other extreme, it becomes a mechanism for confirmation, a tool used to validate decisions already made. The difference between these two outcomes is not determined by software, sample size, or analytical sophistication: It is determined by judgment.

The tension between intuition and analysis is healthy.

The first edition of *The Handbook of Market Research for Life Science Companies* was published in 2017. The book quickly found its audience not because it promised shortcuts, but because it treated market research as a craft that could be learned, practiced, and refined. It was recognized accordingly, earning praise for its clarity and practicality, and it went on to be adopted in graduate programs, incubators, and entrepreneurial settings across disciplines and geographies. For many scientists and first-time founders, it served as a bridge between technical expertise and the unfamiliar terrain of markets, users, and decision-makers.

I met JF around the same time and we hit it off. Even though I came from the "Phelps school of market research," relying more on intuition and frequently glossing over the Total Addressable Market (TAM) pages in pitch decks that came to me as a VC, I found JF's rigor refreshing.

Eight years later, that original achievement remains intact, but the terrain has changed.

The intervening years have not merely introduced new tools or faster workflows; they have altered the conditions under which market research is conducted and interpreted. Certain methods that once felt standard now feel anachronistic. Others have been partially automated, abstracted, or subsumed into platforms that promise efficiency but often conceal their assumptions. And hovering over all of this is the rapid emergence of artificial intelligence—not as a single technique but as a pervasive presence reshaping how information is gathered, synthesized, presented, and consumed.

This second edition should therefore be read not simply as an update but as a re-examination of first principles. What does it mean to "know" a market in an environment saturated with data? How do we distinguish insight from output, signal from noise, and pattern from projection? And what responsibilities follow when market research, particularly in life sciences, does more than inform decisions, but actively shapes which realities are pursued, funded, or abandoned?

One of the most important assertions running through this book is deceptively simple: Market research does not merely describe reality; it participates in creating it. This is not a metaphor. In startups, in academic spinouts, and in early-stage life science ventures, research findings influence which hypotheses are tested, which indications are prioritized, which populations are served, and which paths are deemed viable. Poorly framed research does not just lead to bad decisions; it constructs a distorted version of the world—one that may feel coherent, persuasive, and data-driven, yet remains fundamentally untrue, and one that entrepreneurs then build on and carry forward throughout their entrepreneurship journey. Bad market intelligence does not just lead to one bad decision; it compounds. With each subsequent choice built on the last, stacking one bad decision onto another, the entire company becomes optimized for a reality that does not exist.

Seen from this perspective, the role of market research is epistemological before it is operational. It is about how we know what we claim to know, and how confident we should be in that knowledge.

This epistemological stance has practical consequences, and the structure of this book reflects that. The opening emphasis on the *why* of market research is not introductory filler; it is a safeguard. Too often, research begins with methods before purpose, or with data collection before problem definition. When this happens, even technically competent research can drift toward irrelevance. By insisting that researchers clarify intent, ethical boundaries, and underlying assumptions at the outset, the book places orientation before execution—a sequence that is easy to overlook and difficult to recover once lost.

The chapters on primary and secondary research reinforce this point through practice rather than prescription. Planning, question formulation, and data collection are treated as interpretive acts, not neutral procedures. What is asked, whom it is asked of, and what is left unasked all shape the eventual "findings." Similarly, the distinction between active and passive secondary research highlights a common failure mode: the uncritical reuse of existing information without sufficient interrogation of its provenance, context, or relevance. In an age when access to information is rarely the limiting factor, discernment increasingly is.

This dynamic applies to large established businesses as much as it applies to early-stage startups.

During my early days at Siemens, our market research on PACS technologies gave the usual answers based on replacement of film and film-cost. Eventually, the market size we all agreed on was derived from a critical use case: ease of comparison to past imaging studies for a patient. It has indeed turned out to be correct, but not without many false starts and wrong turns, and overlooking some small, almost irrelevant, sounding elements of the workflow that made adoption start to mirror our market size estimates. One such nit was placing a $20 modem in a $100,000 PACS workstation. Why? The "report had to be faxed to the referring clinician." The modem saved the annoying walk to the printer and the fax machine. None of this is relevant today, but if you place yourself in an imaging center in the year 2000 and take a mental walk through the workflow, the oversight will be quite obvious. The market size for the $100,000 workstation with the silly modem was quite different from the market size without it.

Such discrepancies are just as visible in early-stage startups. Market research answers the general question about market attractiveness quite well. The startups, especially when they are creating a new category, must face a far more specific epistemic problem: Which assumptions about adoption and workflow integration are actually going to help realize their strategy? The proverbial "product–market fit" has a very direct interplay with market research and market size. I can trace back almost all failures in my VC portfolio to being sloppy on this specific challenge.

A few of those startups are long dead or are having a "near life" experience. (Unfortunately), their market size is still correct.

What have I learned from these experiences as a VC? There are only two forms of TAM estimation that I look for: One that has a single word on the slide, "BIG," and one that estimates TAM using the nuances of specific features that allow the TAM to be realized. One page, versus 10 or more pages. There is nothing in between.

A smaller subset of my portfolio companies, on the other hand, managed to unlock progress across most segments in their original market research and their original TAM. The epiphanies in those instances did not come from additional data, but from reframing. We reviewed how questions were posed, what constituted meaningful signals, and which uncertainties truly mattered at that stage of the company's life. The shift was subtle, but consequential. Market research did not suddenly become more sophisticated; it became more honest. That distinction, between accumulation and understanding, runs quietly but insistently through JF's book.

Nowhere is this more evident than in the treatment of data analysis. The book's careful attention to preparation, qualitative and quantitative analysis, and common roadblocks reflects a sober understanding of how insight is actually produced. Analysis is rarely derailed by a lack of tools; more often, it falters due to premature quantification, overconfidence in weak signals, or an inability to reconcile conflicting evidence—and very often, the all-too-human fallacy leading to confirmation bias.

The explicit discussion of the best (and worst) uses of artificial intelligence (AI) in analysis is therefore particularly timely. AI can accelerate pattern recognition, summarize qualitative input at scale, and surface relationships that might otherwise remain hidden. However, it can just as easily amplify spurious correlations or lend false coherence to poorly framed questions. AI is exceptionally good at answering the question it is given; it is indifferent to whether that question was worth asking. Knowing when to trust an output, when to probe it further, and when to discard it altogether remains a matter of human judgment.

Creativity enters the book at just the right juncture. Rather than treating creativity as something reserved for ideation or presentation, this edition embeds it within research methodology itself. The ability to pivot during research, to reformulate questions midstream, or to explore alternative analytical lenses is presented not as a deviation from rigor, but as a response to reality's resistance to neat categorization.

The discussion of analytical frameworks further reinforces this theme. Tools such as SWOT (strengths, weaknesses, opportunities, and threats), TAM (Total Addressable Market)/SAM (Serviceable Addressable Market)/ SOM (Serviceable Obtainable Market), product–market fit, and the Kano model are widely used, often reflexively. Here, they are treated with appropriate respect and restraint. Frameworks are useful precisely because they simplify but simplification carries risk. The value of these models lies not in their mechanical application, but in the questions, they provoke and the assumptions they expose. Used thoughtfully, they clarify thinking; used uncritically, they can harden speculation into facts.

The same attention to judgment carries through to the presentation of findings. Data does not speak for itself; it is always mediated through narrative, structure, and emphasis. The book's focus on the "look and feel" of presentations, storytelling basics, and common pitfalls (particularly for first-time entrepreneurs) acknowledges that the path through which the insights of the entrepreneur reach their audience (customers, partners, investors, employees) is a highly personalized one. The act is not one of transference but one of mapping. The presentation must stimulate the listener, must engage them, and then help them map it back to their own context.

The author's expanded engagement over the past eight years with startups, universities, and incubators shows clearly in this edition. The tone is more pragmatic without becoming reductive; more grounded without losing conceptual clarity. The startup context, in particular, functions as a stress test for market research. Timelines are compressed, resources are constrained, and the cost of self-deception is high. In such environments, methodological shortcuts do not merely reduce rigor; they carry the risk of distorting reality.

I have known the author across these iterations of thought and practice, and I have seen how experience has sharpened, not softened, the core argument of this work. If anything, this edition is more disciplined in its refusal to promise certainty. It acknowledges ambiguity without surrendering to it, and it insists that the ultimate value of market research lies not in prediction, but in understanding. And this leads me to the question: What is understanding, exactly? It requires listening—deep listening. And there is a prerequisite for that: Asking the right questions. My father had a wonderful guide for daily life: Unless you are willing to ask a question whose answer could be a "no," you will never know if the "yes" is actually real. This applies to market research as well. We must ask questions whose answer we do not like. Quoting JF, *"Falling out of love with your ideas is difficult, especially after investing time, money, or reputation."* The antidote lies in having a set of trusted advisors (humans and AI) with whom you develop the comfort zone of discomfort.

The final question posed by the book, "*What is next now that you've created a new reality?*," is therefore not rhetorical. It is a reminder that insight carries responsibility. Once a particular version of the market has been articulated, visualized, and circulated, it begins to exert influence. It shapes strategy, guides investment, and frames success and failure. The ethical task of the market researcher is not to avoid this influence, but to remain aware of it: To revisit assumptions, to remain open to disconfirming evidence, and to resist the quiet slide from inquiry into narrative convenience.

This is why *The Handbook of Market Research for Life Sciences, Second Edition* arrives at an important moment. In an age of accelerating tools and seductive automation, it argues for the centrality of human judgment. It invites readers to treat market research not as a box to be checked, but as an ongoing practice of disciplined curiosity. And it reminds us that in life sciences especially, the realities we construct through research have consequences that extend well beyond the market itself.

Readers who approach this book looking for techniques will find them. Readers who approach it looking for answers may find fewer than they expect. But readers who approach it willing to examine *how* they ask questions, *why* they trust certain conclusions, and *when* to pause before declaring certainty will find something more durable: A way of thinking that remains relevant even as tools and technologies continue to change.

That, ultimately, is the mark of a book worth revisiting . . . and revising.

Ajit Singh, PhD—CEO, Harbinger Health
CEO-Partner, Flagship Pioneering
Adjunct Professor, Stanford University

Introduction to Market Research

Exubera was a dry-powder inhaled insulin product developed to treat diabetes. Even today, it is a captivating example of an innovative product with a strong technical background that might have benefited from stronger market research. It was positioned as a great technical achievement, stabilizing the insulin molecule to make it bioavailable in dry powder form. Furthermore, it was the first insulin product that did not need to be injected, so there were high expectations: Pfizer estimated annual sales would total from $1 billion to $2 billion annually. But those sales never materialized. After nine months, sales of the product barely broke $12 million and captured less than 1% of the market. It was soon removed from the market. In Pfizer's own words, "Pfizer has made this decision because too few patients are taking Exubera."[1]

What Happened?

Blame has been attributed to everything from product management issues to limited advertising, but one thing is sure: There was weak consumer interest for the product, even if it was one of the best clinical products at the time. The reasons for this lack of interest were numerous. For one, users were embarrassed to use the device (it looked like a marijuana bong, was clunky to carry around and to manipulate, and was anything but inconspicuous). Furthermore, the device was difficult to use (you had to carefully measure your dosage), and patients felt that the training was overly complex.

Like most other patients, individuals with diabetes prefer solutions that are easy to use and discreet. Patients with diabetes often feel stigmatized: If you handle a syringe and a vial in a public setting for an insulin injection, you might have the impression that you look like a drug addict, not like somebody who has a chronic disease. Case in point, sensor-augmented insulin pumps—which combine the technology of an insulin pump with a continuous glucose monitoring (CGM) sensor—are now regularly used. These devices are worn directly under clothing, giving patients with diabetes the privacy and convenience they desire.

Since the consumer's perceptions when using the product were not addressed, Pfizer unknowingly had two issues to address to increase user adoption. First, the training for the device had to be optimized for its complexity. Second, Pfizer had to find a way to attenuate the perceived stigma that made the device inherently incompatible with user expectations. It is possible that focus groups with end users or more in-depth interviews with initial users would have identified some of the issues that Pfizer faced or at least given them ways to address them.

This simple example illustrates that companies in life sciences cannot be concerned uniquely with safety and efficacy; they must make sure that they have an accurate read on the market to commercialize their product. It becomes crucial for them to make sure that their product fills a real gap in the market, from both a health and a consumer perspective. Market size is not the only metric of potential success but other elements, such as the competitive landscape, consumer perception, and interests, as well as emerging competing technologies, all play a role in evaluating a potential market.

The objective of this book is to give the reader the perspective needed to organize their market research efforts as well as to develop the tools they need to collect, analyze, model, and present their data. As you will see throughout the book, I believe that involving the customer is important throughout the process, and the earlier the better. When you are developing a new product, it makes sense to involve the customer initially since they will be the ones who ultimately use it, so implicating them early means you are not being developed in a vacuum.

Importance of Market Research

It is not unusual for organizations or founders to believe that they already know their market. Perhaps the decision-makers have been involved in the market for a long time. Perhaps they operated in parallel industries and are making similar assumptions (e.g., what works in industry A works in industry B). Or maybe a key opinion leader (KOL) in the organization is shaping the opinion of other decision-makers, silencing the critical elements necessary to challenge existing assumptions.

But markets are fluid environments. They change rapidly, and yesterday's certainty might already be inaccurate. When things change and new information comes into existence, it is no longer possible to solve current problems with yesterday's solutions. In addition, it is very difficult to convince stakeholders that you know a market based on assumptions. Bringing compelling arguments based on market research is the key to convincing people you have the right idea.

Hence, market research is conducted for two audiences. First, it is conducted for the organization itself to find the information it needs to make effective decisions, to validate market viability and opportunities, and to identify competitive threats and trends. *Market research is conducted when a lack of information will cost the organization more than the cost to acquire the information.*

Second, market research is conducted to convince stakeholders and demonstrate the validity of an organization's business plan elements (e.g., the business model, market entry strategy, product viability, and more). It is used to support (or disprove) internal assessments and beliefs. *The results of market research are a key element of a company's business case.*

As you do market research, you will have to focus and remember that you are looking for the best answer, not simply the best answer you can come up with. This will mean not only conducting a single phase of research but also developing a mindset of continuous and ongoing research.

We break down market research into four components: The process of acquiring the information (which is detailed in Chapters 2 and 3), the analysis of acquired data (which is the topic of Chapter 4), building models for acquired data (Chapter 5), and the presentation of data in a way that is clear and concise (which is explored in Chapter 6).

A Word on Life Sciences

Life sciences is a broad term that is used in many different situations. In this book, life sciences include all industries that have an impact on health (be it human or animal). As such, we include pharmaceuticals, biotechnology, medical technologies, and nutraceuticals. The life sciences industry is characterized by a significant regulatory environment that shapes how companies interact with their customers, purchasers, and end-users. This book will do its best to include this dimension, but fair warning to the reader: Every environment (be it national, provincial, state, or even municipal) might include regulatory and legal components that shape how your product can interact with its end users. As such, just as marketing is an important component of commercializing your technology, so is your overall regulatory environment.

This book will explore methodologies that apply to these sectors of activities. To encompass the most information, we might mention in passing techniques less relevant to life sciences, but our goal here is to zero in on the methods that might be used by startups and growing companies in life sciences, rather than be an all-inclusive repertoire of all that is market research. Whenever possible, life sciences examples (ranging from the hypothetical to past projects detailed throughout this) will be used to illustrate the different concepts. By doing so, it becomes easier for you, the reader to apply these methodologies to your own organization and projects.

One last comment on marketing and life sciences. This is a topic that often is seen with a touch of apprehension. Incorporating a mercantile component into something as critical as saving lives is sometimes perceived as callous and "heartless." However, in the end, you cannot escape the fact that you must be financially viable to provide services and products to those in need, and that without this financial stability, there can be no innovation.

What about Artificial Intelligence and Market Research in Life Science?

As of 2026, it is impossible to write a book that addresses market research and not include artificial intelligence (AI) as one of the key tools. Rather than segmenting it into its own section, AI will be woven throughout the document, detailing the best (and worst) practices. With its rapid progress, there is a distinct risk that many of the suggestions will be dated by the time this book turns 1 year old. Such is the curse of rapid technological developments. In 2023, several healthcare organizations recommended against the use of AI.[2] But today, multiple startups are developing tools that enhance diagnostics, information sharing, and remote monitoring. It is quite impossible to imagine a world where AI is not used daily.

Nonetheless, AI hallucinations must be considered. While AI tools are valuable additions to the market researcher's toolbox, current models can be dangerous in the hands of inexperienced researchers. The current generation of models first attempts to identify existing patterns; when it fails, it silently creates alternative realities, passing them off as the truth. Risk containment in using AI means, for the moment, more human-in-the-loop processes and continuous monitoring of the information generated. Hence, we will be taking a long look through the book on how to use it—and how not to. This will become crucial in developing a critical eye in the use of this technology.

Notes

1 Dariman, T. 2007. *Exubera Inhaled Insulin Discontinued.* Diabetes Self-Management. https://www.diabetesselfmanagement.com/blog/exubera-inhaled-insulin-discontinued/?v=5435c69ed3bc (Accessed January 26, 2026).

2 Dupin, M., & Furkan, O. M. 2025. Enhancing market research with a GPT-based API wrapper: A leap towards advanced data analysis. *Applied Marketing Analytics: The Peer-Reviewed Journal*, 10(4), 348–373. Henry Stewart Publications.

Chapter 1

The Basics of Market Research and Healthcare

There are two types of people in the world. Those who divide everything in two groups, and those who don't.

—Kenneth Boulding

This book is built on a basic premise: Scientific research and market research are not the same thing, especially in a startup. This may sound obvious, but after 25 years of working with scientific entrepreneurs, I can tell you that the word *research* in *market research* creates a lot of expectations.

In science, the researcher seeks precision. They collect large datasets, plan for significant sample sizes, and must make sure experiments can be reproduced. Accuracy matters, and the process is extensive.

Market research in a startup is different; it is closer to detective work. You are trying to understand a social and market reality that has not been uncovered yet. You look for clues, patterns, signals, and trends. You test assumptions and try to map a path that makes commercial sense. You do this with a limited budget. And yes, the sample sizes can be small, and the methods are sometimes crude. Market research in startups rarely gives 100% certainty, but, if conducted correctly, it does give you the confidence to move forward.

If you have ever had to solve a big problem with a restrictive budget, you already know what this feels like. You improvise. You adapt. And that

DOI: 10.4324/ 9781003731726-1

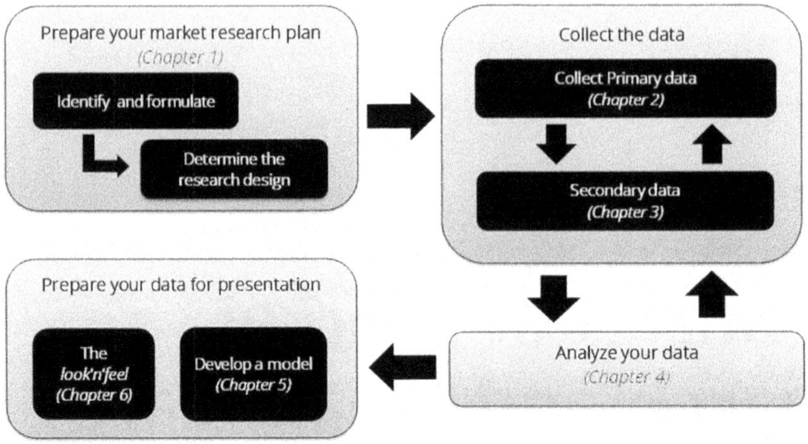

Figure 1.1 The market research process.

resourcefulness is exactly what allows a market researcher in a startup environment to spot opportunities and uncover stories.

Of course, we will be doing this through a systematic approach. We will start by formulating the problem (1.1), then moves to designing (1.2) and carrying out the research (1.3). With data in hand, we will analyze it (1.4) and builds our story (1.5), which we will share with partners, collaborators, and investors. This is the process (Figure 1.1) we will be repeating throughout the book.

However, it is important to remember that life sciences operate in a unique business environment: The healthcare ecosystem. As such, engaging in market research in this space without a basic understanding of it could lead you to misunderstandings or misinterpretations. Complete books exist on the topic, so the goal here is not to be comprehensive. Rather, Section 1.6 is a simple introduction to help you frame and understand basic elements, such as the client and purchasing systems. A more detailed version of this section is available in another book in this series, "*Innovate, Fund, Thrive*" co-authored with my colleague Phillippe Tramoy.

1.1 Formulate the Problem

The first step of the market research process is to clearly identify and formulate the problem (or the opportunity). By formally defining it, you ensure that everyone involved shares the same objectives for the project.

This step usually involves conversations with decision-makers, a review of existing secondary data, and early discussions with key opinion leaders (KOLs). The goal is simple: Align on what you are trying to understand before you start collecting any data.

The topic of research is usually defined in a few words. For example, it could be identifying emerging market opportunities for a new technology, estimating the size and segments of a current market, or developing a customer profile.

1.2 Research Design

The next step is to determine the research design. This is the overall approach you will use to collect your data, and it will guide your choice of the specific methods you will rely on. At this stage, some of the key questions you will need to answer include:

- Which method(s) will I use to collect data?
- How will I connect with my sample? Who do I need to speak with, and what is the best way to reach them? Will I need to incentivize them, and if so, how?
- Which data collection tools will I use (e.g., surveys, interviews, or focus groups)?
- What is my total budget (both in terms of money and time)?

Over the years, with startups, I have increasingly favored using an **iterative market research approach.** The concept is simple: Small research steps guide you toward the next phase. You do not need to launch a 2,000-respondent survey on day one. In fact, you might get far better insights from three or four small surveys as you gradually get closer to your true target audience and ideal customer. Your research evolves as your understanding evolves.

This is very similar to **agile marketing,** where constant iterations are supported by continuous feedback. You do not let one customer define the entire focus, but you do integrate what you learn into your next step. For example, you might run five interviews, pause to reflect, and then perform another five based on what you discovered in the first batch. It is about doing more with less and doing it efficiently.

Too much research early on can slow you down, where you feel locked into your initial assumptions simply because you invested heavily in a large, complex study. When that happens, founders sometimes hold on to those first results even when newer data shows a different reality. Iterative research mitigates this risk, as your perspective constantly evolves through your research.

1.2.1 Miles-Wide/Inch-Deep Versus Deep-Dive Research

When approaching a market research project, a researcher must decide whether they are going for a "miles-wide/inch-deep" approach or a deep-dive approach.

The miles-wide/inch-deep approach focuses on getting a broad overview of a segment, industry, or competitive landscape. It is used early in the process, when you do not know what you do not know. The goal is to gather information across as many data points as possible simultaneously. By its nature, this type of research is exploratory. At the conclusion of the project, you should have a clear, high-level view of what you were researching.

The deep-dive approach is different. It is used to validate the best opportunities identified in the miles-wide step. It involves concentrating on one specific, predefined topic. In this case, the top trends, competitors, or issues have already been identified, and the researcher invests their energy in understanding each one in depth. Instead of interviewing generalists, you meet with KOLs. Instead of general reports, you consult specialized resources.

Many research projects are a combination of both approaches: You start with a miles-wide/inch-deep scan to understand the landscape and then conduct a deep-dive into the most promising or critical elements. The key is to align expectations with the resources you have.

1.3 Data Collection

Next comes your data collection. This is an essential step, and there is one rule of thumb that sits above all others: Never trust data unless you know where it comes from and how it was gathered. This is especially important when dealing with secondary data, where you may not always see the underlying methods used for data collection. If you do not validate the source, you risk building your strategy on external mistakes or misconceptions.

There are two main types of data you can collect: **Qualitative** and **quantitative.** You will also need to decide whether to rely on **primary** or **secondary** data. Finally, you will need to develop a **sampling plan** that outlines who you will reach out to and how.

1.3.1 Quantitative and Qualitative Data

Data can be either **quantitative** or **qualitative.**

Quantitative data refers to information that can be counted or measured. Examples include the number of potential customers for a product, how many doses of a drug a patient uses daily, or the average distance someone is willing to travel to visit a specialized clinic.

People in life sciences are usually comfortable with quantitative data because they already rely on it to assess technical performance. In market research, quantitative data is often used to size markets, identify segments, and evaluate opportunities.

Later in this book, we will see that some data collection tools are better suited to quantitative work. Surveys (whether online or in person) are typically the most efficient way to collect large quantities of quantitative data.

Qualitative data is subjective and interpretive. It can include stories, opinions, observations, pictures, and even audio recordings. Examples include the personal reasons behind product preferences, the perceived impact of quality on purchasing decisions, or how packaging colors influence buyer behavior. Interviews, focus groups, and observational studies usually produce qualitative data. And as we will see later, qualitative responses can sometimes be coded and transformed into quantitative insights.

In my experience, academics and investors usually prefer quantitative data. For startups, however, generating large sets of quantitative data can be difficult, especially in emerging or highly innovative markets. Tables, pie charts, and frequency counts are easy to present, and many individuals instinctively view them as "real data." However, qualitative data plays a crucial role: It provides color, context, and meaning to the numbers.

While qualitative data is essential, it is also one of the most misunderstood forms of data in early-stage startups. Founders often over-trust the directionality of a single strong anecdote.

For example, I once worked with a startup that had positioned their product firmly in the education market after meeting one overenthusiastic teacher. Two years later, as their prototype was taking shape. I was brought into complete market research and develop a commercialization plan. As I was conducting interviews with potential clients (over 20 interviews) it became clear that the education market was a poor fit for their product, and that they had a business-to-consumer (B2C) play. This was a significant shift, especially since they had spent 2 years collecting data in schools and classrooms that no longer supported the best commercial opportunity. In the end, individual stories are useful, but they are not the way to define a market.

In another project, I worked on a project with an Australian advertising agency preparing a promotional campaign for a major pharmaceutical company. We surveyed more than 3,000 consumers on their use of painkillers. The quantitative data indicated which brands were preferred but it did not explain *why* the leading brand was preferred, especially since my customer perceived it as weaker and less effective.

Only after analyzing the qualitative responses did the patterns emerge: Family habits, health concerns, routines, price promotions, and advertising all played key roles in shaping preferences. To deepen the analysis, we monitored spontaneous online discussions. That revealed passionate debates regarding side effects, home brands versus traditional brands, and highly specific perceptions concerning strength versus digestive tolerance. The result? The number one brand was not viewed as "stronger," but it was viewed as *safer* and more *familiar*, and as such was more frequently recommended. Numbers alone could not have revealed that.

The key lesson here is that convergence of data sources matters. Quantitative data is often perceived as more concrete and is easier to present, while qualitative data adds depth, context, and meaning. Strong market research usually requires a convergence of both.

1.3.2 Primary and Secondary Market Research

There are two types of market research you can engage in: Primary and secondary market research.

Primary market research consists of activities where the researcher generates original new data. This can include web surveys, in-depth interviews, focus groups, or any method where you collect information directly from participants. In other words, this data did not exist until you generated it. It is tailored to the specific needs of your project. We will explore primary market research in detail in **Chapter 2.**

Secondary market research involves data that already exists and must be collected, reviewed, and organized. This can originate from web searches, published reports, databases, academic papers, or even aggregated consumer comments, posts, and blogs. The role of the researcher is to gather this existing information and transform it into something coherent and useful. Secondary market research is covered in **Chapter 3.**

1.3.3 Designing a Sampling Plan

A sampling plan is the detailed framework that outlines who will be contacted and what the expected sample size should be. Sample size is crucial whenever the validity of the information you collect matters, which is why estimating it beforehand is essential. There are two main types of samples a researcher can use: **non-probabilistic** and **probabilistic.**

Non-probabilistic samples are those in which participants are *not* selected at random. They are frequently used when dealing with experts, niche populations, or addressing practical constraints. Common types include:

- **Convenience sample** where participants are selected because they are available and willing to take part. For example, if your company has a kiosk at a trade show, collecting responses from attendees would constitute a convenience sample. This sample is often easier to collect but comes with a huge potential risk: Unknowingly, the researcher might be inserting a bias in his research (e.g., geographic bias or professional network). This type of sample can be used to start the research process but its data should be used sparingly for later decision-making.
- When using a **quota sample,** the researcher sets predefined requirements for the composition of the sample. For instance, he makes sure that at least 50% of respondents are women or that 40% come from a specific geographic region.
- As for the **snowball sample,** it is used when participants are highly specialized or difficult to reach. You begin by interviewing KOLs and ask them to recommend others. The sample grows as participants refer additional individuals. When the chain stops producing new leads, you may need to restart recruitment from a new entry point.

- Finally, for a **voluntary sample,** it occurs when members of a defined group choose to participate. For example, in an association's annual survey, the respondents form a voluntary sample.

As for probabilistic samples, they rely on some form of random selection. This approach is more statistically rigorous and is commonly used for descriptive or quantitative research. There are three types of probabilistic sampling:

1. *Simple random sampling:* This method is used when every individual in the population has an equal chance of being selected.
2. *Stratified sampling:* This approach is used when the population is divided into homogeneous subgroups (strata), and a simple random sample is taken from each subgroup.
3. *Systematic random sampling:* This technique is used when participants are selected according to a random starting point and a fixed interval (e.g., surveying every third house on a street).

Let's take a moment to discuss how many people you should contact. It is one of the most common questions in any market research project. Unfortunately, there is no universal rule of thumb. Before determining the sample size, it is essential to conduct exploratory research, which helps approximate the size of the user base or potential market. Once you have this estimate, you can use a basic statistical model, (or simply use an online sample size calculator) to determine a valid sample.

Of course, budget and time constraints play a major role. While interviewing 500 people individually might be ideal from a scientific standpoint, it is rarely feasible in a startup environment. In practice, even 20 well-chosen interviews can generate a lot of insights. A recent article suggested that approximately 30 interviews are sufficient when the goal is to identify subgroups during analysis.

Overall, I usually use a simple rule of thumb: Once you begin hearing the same information repeatedly, you are reaching saturation, which means your sample is likely adequate.

Of course, whenever data gathering is inexpensive, aim for the maximum sample size you can collect with the resources at hand. For example, if you have access to multiple individuals or potential clients willing to participate without requiring an honorarium, it is strategic to interview as many participants as possible.

1.3.4 Design Your Tool

Once the research question has been defined and the methodology selected, it is time to design the research tool. For example, if you plan to conduct interviews, you will need to build an interview guide. A good guide ensures consistency from one interview to the next, across different interviewers, and acts as a helpful prompt to make sure all key topics are covered.

Ideally, you should test your research instrument before using it widely. Try your interview guide with a few potential participants: You may discover that some questions are redundant, others are missing, and some are misunderstood by your target audience. It is far more cost-effective to identify these issues now rather than during data analysis.

In market research, the two main types of tools you will prepare are questionnaires and research guides (including variants such as discussion and observation guides).

A questionnaire is a structured list of questions used to collect data from a statistically significant number of respondents. It is most frequently used in online surveys, in-person surveys, telephone interviews, or mailed questionnaires. In many instances, the market researcher will *not* be administering the questionnaire themselves: it may be deployed through an online platform or carried out by a third party. This means the questions must be clear, unambiguous, and standardized across all interviewers or platforms.

A discussion guide, on the other hand, is typically used for focus groups and in-depth interviews. It relies mainly on open-ended questions that guide the conversation while allowing participants to expand on their thoughts.

More information on building these tools is available in **Chapter 2.**

1.3.5 Collecting Data

This is the moment when you head into the trenches and put your plan into action. You begin recruiting and conducting interviews, you send out your survey, or you start searching the internet for the information you need. This work can be conducted by you, or you may hire a firm to handle some (or all) of the data collection.

We will cover this step-in detail in **Chapters 2 and 3** but the key point to remember is that data collection is often the most time-consuming part of the market research process.

1.4 Analyze Your Data

Once you reach the end of your data collection, it is time to start analyzing the results.

If you have **quantitative** data, you can use a simple spreadsheet software (e.g., Excel) or statistical software (e.g., SPSS or JMP). You will build tables and graphs, often correlated with demographic variables such as age, geography, or profession. From there, you will start looking for trends and patterns.

If your data is **qualitative,** the first step is transcription and codification. Codifying responses allows you to identify consistent patterns in what people say. You might begin by creating broad categories, grouping similar responses together into meaningful buckets. In some projects, it can be helpful for two people to code the data independently and then compare findings to reduce bias.

The use of artificial intelligence (AI) cannot be ignored when discussing data analysis. Indeed, AI greatly enhances a researcher's ability to process and analyze large amounts of data. It can ease the burden of manual transcription and reduce miscalculations due to human error. Automated searches, auto coding, and integrated analysis tools all facilitate researcher tasks. It also facilitates the analysis of large datasets, pulling together patterns which would take an individual researcher enormous amounts of energy. That is why discussion on best-case uses will be at the center of the chapter on data analysis.

Of course, avoiding analysis paralysis risk is an important element. Founders often keep analyzing instead of deciding, wait for just one more data point, or chase statistical perfection in an early-stage context where it does not matter. When patterns stabilize and diminishing returns kick in, it is time to decide.

We will explore different methods for making sense of both quantitative and qualitative data in **Chapter 4.**

1.5 Storytelling

Once you have interpreted your data and transformed it from raw information into a coherent story, it is time to share it with stakeholders, investors, and potential partners. You may want to spend a few minutes presenting the methodology you used to collect the data—this helps reinforce the validity of your findings—but the bulk of your presentation should focus on the results, the insights that emerged, and the recommendations that follow from them. And as we will explore, the story does not stay the same for everyone; investors, partners, and internal teams each need their own version of it. I often tell the startups I work with that, by the end of your journey, you will probably have a folder on your network with more than 50 slide decks, each customized for a specific audience or investor.

This part of the process is so important that two full chapters are dedicated to it. **Chapter 5** will cover different ways to present data and how to classify it in a coherent way to generate additional insights. **Chapter 6** will focus on the "look and feel:' How to present your data in a clear, consistent, and punchy manner and, just as importantly, what *not* to do.

1.6 Overview of the Healthcare Ecosystem

The traditional healthcare ecosystem is usually comprised of four main components.[1] At its core, we have the individual patient, who is taken care of by a care team. This team includes professional care providers (e.g., clinicians, pharmacists, and other healthcare professionals) as well as family members. Encompassing them, we have the organization that supports care (e.g., hospitals, clinics, and nursing homes), which itself is impacted by a political and economic environment (e.g., regulatory frameworks, financial systems, payment regimes, and market dynamics). These define the conditions under which organizations, care teams, individual patients, and individual care providers operate. This is illustrated in Figure 1.2.

In our first bubble, we have **the patient,** who is (hopefully) at the center of the healthcare system. In many cases, the patient is the end-user of a product and, as such, is often perceived by innovators as a client. Hence, product development and marketing strategies are built with the patient in mind, but he might have only a very limited input into purchasing decisions.

On the next level, we have **primary caregivers** (e.g., physicians and nurses), who play a vital role in adopting new products in a healthcare setting. These stakeholders can be positioned as champions for technology, being asked by the facilities to share insight on which products are worthy of consideration. However, doctors are not always decision-makers in hospital settings, so while a product that makes their job easier might

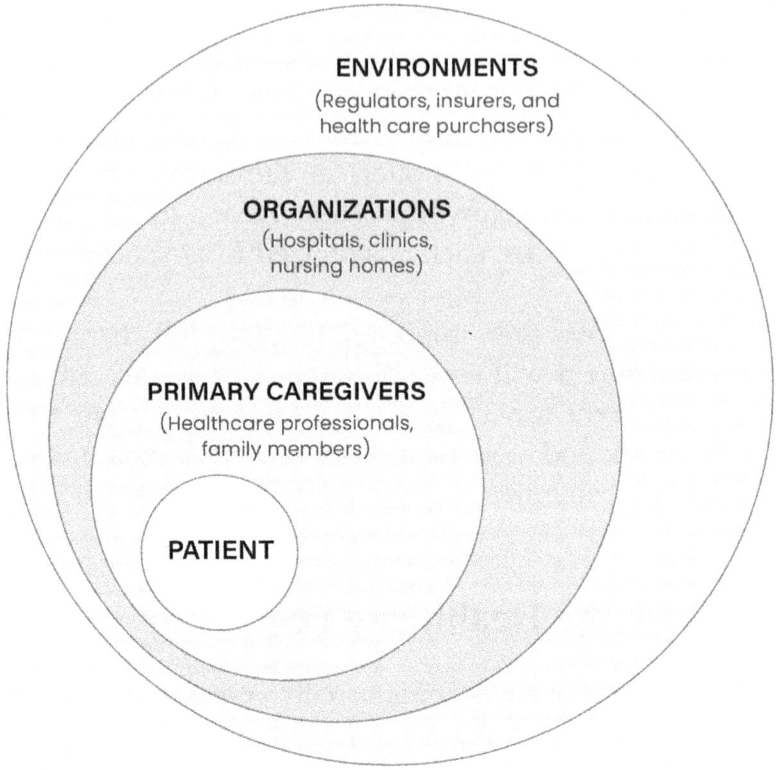

Figure 1.2 Illustration of the traditional healthcare system.

interest them, if it is costly compared to the current gold standard, it could be difficult to gain traction.

In this second level, we also have the **primary caretaker,** who, *while not always implicated in the healthcare decision,* can be an important influencer in the purchasing decision. A caregiver is usually defined as someone from the patient's network who helps them with daily activities.

In our third level, we have the infrastructure that supports healthcare. **Hospitals** are organizations dedicated to giving care to patients in need. They can be for-profit or nonprofit. Depending on their relationship with Group Purchasing Organizations (GPOs) and insurance companies, they can have various levels of latitude when selecting products.

Also included in the infrastructure are clinics and nursing homes. **Private clinics** usually have more flexibility around purchasing innovative technologies and are often targeted as first clients for innovations, as they are often better equipped for small-scale integration and can be leveraged as successful success studies. As for **nursing homes,** they usually have limited budgets for innovation but will be receptive to technologies with potential for cutting costs or improving safety. On the last level, we have the group environment around innovation, which includes regulators, insurers, and GPOs.

Regulators play a significant role in defining the rules around the delivery of healthcare (what can and cannot be conducted) and often interact with companies when evaluating the safety of the devices being developed. **Private and public insurers** play a significant role in healthcare as, they reimburse part or the totality of a treatment, product, or device they have specifically selected to cover. **Group purchasing organizations** are organizations combining several like-minded healthcare organizations, leveraging a combined group purchasing power to lower the price of a product or service due to higher expected volumes.

The framework we just reviewed has several advantages. First, it puts the patient at the center of the intervention. It is also simple to understand and enables a quick and easy understanding of how healthcare works. It also highlights a basic model of how each actor is impacted by those above it.

1.7 Concluding Word: Basics of Market Research

In the end, market research is simply a structured way to reduce uncertainty and make smarter decisions. My objective in this book is to share my experiences and insights so that you, the reader, may be able to engage in this journey yourself and find the insights you need to create a successful venture.

Note

1 National Academy of Engineering (US) and Institute of Medicine (US) Committee on Engineering and the Health Care System; Reid, P. P., Compton, W. D., Grossman, J. H., et al., editors. 2005. *Building a Better Delivery System: A New Engineering/Health Care Partnership.* Washington (DC): National Academies Press (US); 2, *A Framework for a Systems Approach to Health Care Delivery.* https://www.ncbi.nlm.nih.gov/books/NBK22878/ (Last visited January 28, 2026).

Chapter 2

Primary Research

Enable a curious mindset. Curiosity encourages time and headspace for asking questions, which can lead to new insights.

—Caroline Florence

Primary data is information that is generated directly by the market researcher to answer their research question. For example, when they are conducting interviews, online surveys, or making observations, they are gathering primary data.

Generally, primary research is costlier to generate (both in terms of time and resources) but it is customized for the researcher's needs. If they have correctly designed their tools, they should be able to solve their research problems. Also, the data they collect is proprietary, so it belongs to the organization exclusively, becoming a competitive advantage.

This chapter is divided into two distinct sections. The first section deals specifically with the data collection tool (i.e., the questionnaire or discussion guide), while the second section deals with the information collection activities specifically—such as in-depth interviews (IDIs), focus groups, surveys, observation, mystery shops, and Delphi groups—from an implementation perspective and provides examples and information on how these activities are used by life science organizations.

DOI: 10.4324/ 9781003731726-2

2.1 Importance of Preparing a Market Research Tool

Properly preparing your market research tools is important as it ensures the quality of the data collected. It does so by ensuring that data is collected in a consistent manner and the topic is covered comprehensively.

First, data needs to be collected in a *consistent manner*. To be able to compile data from different investigators or from different time periods, it must be collected in a consistent manner. The questions must remain the same from one sample to the next. It is problematic to compile data if some respondents answered questions that were phrased differently. This is especially problematic in surveys where data is collected in multiple languages. Extra time in the testing and validation phase must be taken to ensure survey computability.

For example, a few years back, I was brought into a project to analyze some primary data a client had collected, as they were unable to build consistent models. A web survey had been posted on two different websites that belonged to two different subsidiaries. Each subsidiary had designed a different autonomous web survey (based on the initial questions) and had collected the data independently. After a careful review, I noticed some subtle yet significant differences between the questions in the two surveys. To improve demographic data, question and answer choices were changed in one of the two surveys. The changes were significant enough that the online survey was not collecting the data the same way in the two locations. As such, it was impossible to merge the data from the two sources, and the data was impossible to tabulate correctly until some of the data was recoded in these surveys.

Also, data needs to be collected in a *comprehensive manner*. During an interview, it is easy to get wrapped up in the conversation and forget to ask questions on a specific topic or skip them entirely. Having a well-designed tool can help the interviewer remember all the topics they wanted to cover. If you took time to prepare questions on a topic, it is quite likely to be important. While it is occasionally possible to go back and ask the person interviewed, the dynamic created in the initial interview is lost.

2.2 Designing a Data Collection Tool: Step by Step

Building your data collection tool is an important step. If possible, multiple individuals should be involved at various stages to ensure that all the information needed is collected. I propose here four simple steps to develop your data collection tool.

2.2.1 Step One: Define the Context

As discussed earlier, the first step in conducting market research is to define the information required, the target respondents, and the data collection tool you will use.

- *Defining the information required* helps the researcher focus their questions. While they might have defined their needs from a broad perspective, it might be useful to conduct a little secondary research first to gain some additional insight and help clarify what information really needs to be created versus what already exists.
- *Next, identify the target respondents.* Defining them clearly helps conceptualize the knowledge level of the respondent and their technical expertise, which influences the way the questions are going to be prepared.
- *Finally, the choice of the data collection tool* will dictate how the questions will be asked and worded. For example, a questionnaire might be self-administered (the participant is alone when answering the questions) or assisted (with an interviewer asking the questions and clarifying topics if needed). If it is self-administered, you should consider that the user will have to interpret the questions and answers by themselves, and more detailed questions might be necessary.

2.2.2 Step Two: Build Your Question Bank

A brainstorming session with the parties involved is a great way to start building your question bank. During this creativity phase, the objective is to identify all the topics you want covered by your market research activity, to generate as many questions as you think you will need to ask, and to try to cover as much ground as possible. You can weed out duplicates and unnecessary questions when you build your data collection tool. We go into more detail on how to design questions in Section 2.3.

2.2.3 Step Three: Build Your Data Collection Tool

Once you believe you have generated enough questions to cover all the ground you need covered, it is time to place your questions in a meaningful order.

a. *First section—Introduction and breaking the ice*
   - Start your data collection tool with a brief statement explaining why you are collecting the information and, if applicable, stating that the information is entirely confidential and that only consolidated results will be published.
   - Participants might be hesitant to answer questions and share sensitive opinions if they do not know what the information is going to be used for. Clearly stating why you are collecting the data will contextualize the market research activity for the participants and increase participation.
     - Reminding users at the start of the data collection tool that the information will not be shared at large increases participation and ensures the participants share more truthfully, while mentioning that it will only be shared in a consolidated form makes it more reassuring for participants.
     - Include warm-up questions: Start your question guide with a few simple questions to engage the participants and ensure that they feel comfortable in the process. If you are using a web-based platform, including a few easy warm-up questions lets the user gain familiarity with the online survey platform. Some simple warm-up questions could include questions about the person's job title and function, for example, or their relationship to the topic covered in the data collection tool.

b. *Second section—Core of the data-collection tool*
   - Put your most important questions in the core of your data collection tool. Do not put the most important questions at the end of the question guide, as you do not want your participants suffering from participant fatigue when they get to them, as they may botch or skip them to finish the survey.

- Place the most important items in the first half of the questionnaire to increase the response rate on the important topics. If you are short on completed surveys; you might be able to use partially completed surveys to fill the gaps.

- Make sure the questions flow from one to another and do not force the participants to go back and forth on topics. If you ask a series of questions on how they use a medical device, try to group these questions together to keep participants in the same mindset.
- When building your data collection tool, be critical of your questions. Ask yourself: "Is this question really useful?" Is it "need to know" or "nice to know?" Participant attrition occurs when data collection is too long, so make sure you only keep the important questions you need answered.
- Include some variety in the type of questions you ask. Too many open-ended or closed-ended questions clustered together might bore participants. Instead, mix the type of questions you ask to sustain user participation.

c. *Last section—Wrapping-up*

- Remember that as the participant nears the end of the data collection tool, they might become increasingly indifferent, giving careless or half-thought-out answers (as they are just trying to reach the end of the survey to get their incentive).
- Questions that are very sensitive in nature can be included at the end of the survey to avoid losing participants before other important information is collected.
  - For example, asking a question on a sensitive topic (e.g., assisted suicide, abortion) might deject participants. Placing the question at the end lets you use the other data they provided before dropping out.
- One wrap-up question I have taken a habit of adding to all my interview guides is the message to the founders or company. Something akin to:

> "We just finished talking about this innovation. Lots of great insight. As we wrap-up, if you could talk to the entrepreneur directly, what would you say to them? What is the most important element from today's discussion you feel they should know?"

This forces a reflection and gives a chance for the participant to synthetize an hour's worth of discussion into what they believe is the key learning.

- At the end of the survey, remember to include questions related to demographics. This will ensure that you can compare the answers between segments, as well as correlate answers and identify trends.
  - Some researchers prefer putting demographic questions at the beginning of the survey instead, using them as warm-up questions. That is acceptable and especially recommended if you have a few sensitive questions at the end of your questionnaire or feel that participant attrition will be high.

2.2.4 Step Four: Validation

Before deploying your data collection tool at large, test it with a sample of your target population. If your objective is to interview end users, try your questionnaire beforehand on a few end users to identify any gaps or misunderstandings your participants may have. This helps ensure that the questions are clearly asked and that the data you collect is what you need, as well as being easy to compile and interpret. It is also useful to make sure that the questions are answered consistently across different individuals. Finally, it is useful to validate your survey so that you can remove questions that are not really needed or that are redundant.

Pre-testing your data collection tool also lets you validate its length. If you see participant fatigue during the testing phase, consider removing some questions or adjusting them from open- to closed-ended. As a rule of thumb, an online survey should take about 20 minutes to fill out (unless you are offering an incentive or you have an especially captive audience), while an IDI could reasonably last 30–45 minutes and a focus group could last between 60 and 90 minutes. Of course, if you are targeting participants who are very busy, a shorter tool is needed: A project targeting chief executive officers (CEOs) should include only four to five key questions to ensure a maximum participation rate.

After testing, you should be ready to start data collection. You might be able to salvage some of this test data for analysis; if you change 2 questions, and keep 10 questions the same, the data from those 10 questions could be used during the analysis.

2.3 Formulating Questions

In this section, we will go over some of the finer points relating to formulating questions. Take note that the golden rule is to *keep your questions simple.* The more complex the question is, the more likely the person answering it will either skip it, not answer truthfully, or just answer incorrectly as they do not understand what you are trying to ask. This section will deal with both closed-ended questions and open-ended questions.

2.3.1 Closed-Ended Questions

There are three types of closed-ended questions: Nominal, ordinal, and interval questions.

a. *Nominal questions* are those where the choices are presented to participants across categories but in no order. Asking users which color they prefer or which brand of painkiller they use is a nominal question.

   An important side note here is that the order in which options are presented to a respondent can influence their answer choices. This is called the primacy effect: In paper and internet surveys, respondents tend to pick the first choice of a list rather than reading through every available option. If you are using an online survey tool, use a random sorting option to attenuate the primacy effect.

b. *Ordinal questions* are used when there are different answer categories with an importance ranking across possible answers. For example, a market researcher may ask to rank the performance of a product across a scale of slow, medium, and fast.

   It is important to make sure that there is no overlap or confusing terms across categories. For example, a scale of "small, average, large, big" possesses several issues: What is average? What is the difference between large and big? A much better scale would simply be "small, medium, and large."

c. An *interval question* is one where the spacing across categories is even. For example, asking users how much they like something on a scale of 1–7, where 1 is "not at all" and 7 is "the most," is an interval question. A good interval scale will usually have five to seven points and will

Table 2.1 Sample Likert Scales

For Measuri`ng Agreement	*For Measuring Frequency*	*For Measuring Likelihood*
• Strongly agree • Agree • Undecided • Disagree • Strongly disagree	• Always • Very frequently • Frequently • Occasionally • Rarely • Very rarely • Never	• Definitely • Very probably • Probably • Probably not • Definitely not
For Measuring Importance	*For Measuring Quality*	*For Measuring Value*
• Very important • Important • Moderately important • Slightly important • Not important	• Very good • Good • Acceptable • Poor • Very poor	• High • Moderate • Low • None

include a middle category to produce better data, as this lets people who evaluate the item being measured as average answer truthfully. Indeed, it is interesting to consider that psychologists have been using seven-point rating scales for a long time, on the basis that increasing them into finer categories does not really add much to the usefulness of the ratings.[1] This is called a Likert scale. As a point of reference, Table 2.1 demonstrates a few standard Likert scales to help you in your questionnaire building.

Remember that points on the scale should be labeled with clear, unambiguous words. Also, it is important to be consistent in building scales throughout the questionnaire: Do not go from low to high in one question and then switch from high to low in the next one. Finally, building items that are symmetric (meaning they have the same number of positive and negative items) ensures a neutral data collection tool.

2.3.2 Open-Ended Questions

Open-ended questions are questions that require explanation from the participant. They require the participant to engage and justify their position, while closed-ended questions are questions where the participants have only a few possible answers. Consider the following two examples relating to headache medication:

- Question #1: Enumerate all the different brands of headache medication you know. (Open-ended)
- Question #2: Do you know the headache medication [XYZ brand] (yes/no)? (Closed-ended)

In the first question, the researcher can generate a significant amount of information, such as:

- The consumer's knowledge of different brands
- Which brand or brands come to mind?
- Which one came to mind first?
- The order in which they came to mind
- Which one or ones the consumer does not have any awareness about

In the second question, the researcher is learning exclusively about the consumer's knowledge (or lack thereof) of the [XYZ] brand.

If there are many people to question, the data from open-ended questions will need considerably more resources to be analyzed and codified, while data from closed-ended questions can be analyzed quite rapidly. Hence, budgetary concerns weigh in when choosing between open- and closed-ended questions.

The other issue with open-ended questions is that they require more effort from the participant, giving them more "room to act" and answer falsely. Some participants take the opportunity provided by an open-ended question to share displeasure with the product, the company, or even the survey itself, even if it has no relevance to the question. You can easily have 5% of responses in open-ended questions that are incoherent and must reject them.

2.3.3 Projective Questions

Projective question techniques are used when the market researcher wants to get more insight into a participant's perception of a brand or product. Usually, participants are asked to project their thoughts onto other things and are then asked to explain their answers. This probing into the participant's thought process can provide more insight into the reasons they made these associations. Common projection techniques include:

a. *Word association questions:* For this type of question, participants are given a list of words (or a series of pictures) and are asked to respond with what comes to mind. This provides the researcher with insight into the consumer's vocabulary in relation to a brand or product. By using the results of multiple answers, researchers can uncover market trends.
b. *Sentence and story completion:* This involves asking a participant to complete a sentence or story. This helps to gain insight into the participants' predetermined attitudes and thoughts. It is mostly used when direct questioning is unlikely to provide an honest response (e.g., the participant might be too shy to answer truthfully). This type of question is useful in brand positioning and commercialization. For example:
   - People who use Brand XYZ are ____________
   - I only buy product XYZ when ____________
c. *Photo sort questions:* This technique involves giving participants a series of photos and asking them to select which ones they associate with a product or brand. Afterwards, the participants are asked to explain why they chose these pictures to gain additional insight. This can very useful when identifying a product's perceived positioning, such as for a medical device or laboratory product.
d. *Third-person technique:* This technique involves asking participants what they believe other people are thinking, feeling, and saying. This is useful when tackling subjects that are sensitive in nature, and you expect people might hide their true feelings on a topic. Topics in healthcare that are highly sensitive (e.g., assisted suicide or abortion) are often best explored using the third-person technique.

There are other techniques used in market research that are less common in life sciences, such as:

- *Cartoon drawing:* Presenting participants with a cartoon and asking for their interpretation.
- *Personification:* Turning a brand into a person and describing that person's characteristics.
- *Stereotyping:* Presenting a description of a specific group of people and asking questions about a situation of interest.

2.3.4 Choice Modeling Questions

If you are trying to ascertain an individual's decision process, it can be a good idea to include choice modeling questions.

Choice modeling consists of asking a series of discrete questions where the participant makes relative choices (e.g., A vs. B, B vs. C, and A vs. C) allowing the researcher to infer the prioritization between A, B, and C.

To build a choice model, you must properly identify the service or product to evaluate and then select the attributes you are testing. For example, if you are creating a portable diagnostic device, you might be interested in learning whether speed, accuracy, price, robustness, or portability is the most important attribute. By asking a series of questions opposing one attribute to another, you can identify a pattern. It is possible to build more complex models, but for market research purposes, a simple model can often provide a wealth of information.

The advantage of choice modeling is that it forces participants to make choices. Asking if speed, accuracy, price, robustness, and portability are important will most likely leave you with inadequate data (they are all important after all). However, a series of questions that force the user to make a decision is more likely to encourage the consumer to reflect and can generate a more definitive model. It can also be used to estimate the impact of pricing on specific attributes and reduces the participants' ability to bias the research results.

The main disadvantage is that the data generated is ordinal; therefore, it provides less information than ratio or interval data. So, you might learn which attribute is the most interesting but will not necessarily know how much more important it is.

Overall, the primary use of choice modeling is for predicting preferences, refining new product development, estimating the willingness of customers to pay for goods and services, estimating the impact of product characteristics on potential consumer purchases, and evaluating the importance of product attributes.

2.3.5 Question Banks

Question banks can be a great source of inspiration when building a questionnaire. Some sources for preexisting questions are shown in Table 2.2.

Table 2.2 Online Websites Featuring Banks of Preexisting Questions

Name (Website)	*Comments*
The General Social Survey (https://gss.norc.org/)	Contains a core of demographic, behavioral, and attitudinal questions, plus topics of special interest.
American National Election Studies (ANES) (www.electionstudies.org)	ANES produces high-quality data from its own surveys on voting, public opinion, and political participation.
Association of American Medical Colleges (www.aamc.org)	Some useful question examples from surveys conducted in the last few years.

2.3.6 Mindful Surveys

A different approach to designing survey questions is the Mindful Surveys Technique.[2] It integrates qualitative and quantitative questions into a two-step process to build an iterative questionnaire. This is especially useful for companies with a limited budget or for researchers with limited knowledge of the topic that will be investigated.

As mentioned earlier, analyzing a large qualitative data set can be time-consuming and costly (especially if you must conduct a double-blind evaluation by two independent researchers). A way to mitigate this is to use closed-ended answers; however, you might not feel confident enough to enumerate all possible choices. In this situation, you might prefer to collect more information before developing a final closed-ended question.

Mindful surveying is a two-step process to resolve this issue. The first step involves opening the survey to participants using open-ended questions and collecting approximately 100 responses. Once a sufficient number of completed surveys are received, the researcher temporarily closes the survey and analyzes the data. Using those responses, the researcher can identify patterns and trends to craft a closed-ended question. He can then modify the survey with his new question, open the survey again, and collect the data in a more digestible format. If he could not identify enough trends, he could open the survey again, collect another hundred entries, and analyze and try to identify trends again.

The researcher should remember when using this approach that it is important to set guidelines on the "stop-go sample" so that specific groups are not underrepresented. For example, while collecting 100 responses could be an end point, if the researcher is surveying different age groups, it may be more relevant to stop the survey only once 50 answers are received for each age group.

2.3.7 Things to Watch for

Here are a few items that one should watch for when formulating questions:

a. *Double-barreled questions:* A double-*barreled* question is a question that attempts to measure multiple items simultaneously. This makes it difficult for participants to answer and leads to incorrect or unusable data. For example:
   - "*What is the quickest and most precise test to verify XYZ?*"
   - Are you interested in the quickest or the most precise test? What if one test is the quickest and another is more precise . . . which one should the participant choose? Respondents who believe one test is the quickest and another is more precise would be unable to provide an accurate answer. If they do respond, the data supplied would likely be a compromise. Instead, the correct solution is to split the question into two distinct questions, each having its own specific objective.

b. *Avoid guiding your users towards a response:* While it would be convenient to obtain a strong data set that confirms market intent for the next steps of our development journey (e.g., fundraising, marketing strategy, and so on), intentionally guiding users toward a specific answer is not conducive to an objective market report. As such, questions such as:
   a. Do you like medical devices that make you feel better?
   b. Do you like when medication solves an ache quickly?

   Might be great to illustrate needs in the market, but do not really reflect anything significant, as all consumers tend to favor technology that heal them. Phrase your question to find useful data, even if it means finding out your innovation might not be as useful as you initially thought.

c. *Choosing the right words:* One of the biggest issues is the choice of words. Incorrect words can lead to bias, rendering data useless. Consider the following question:
   - Do you like it when your medication relieves pain quickly?
     Very few people do not like quick pain relief, and the question as worded leads the person to answer in a highly specific way. A more effective question would be:
   - What characteristics do you look for in pain medication?

d. *Overlapping interval answer sets:* Overlapping interval answer sets occur when response categories are incorrectly designed, causing one or more categories to overlap. Consider the following question:
   - How many times do you see your doctor every year?
      i. *Never*
      ii. *1–3 times a year*
      iii. *3–10 times a year*
      iv. *10 times a year or more*

   See the issue? The categories overlap, so it would be impossible for someone who sees a doctor three times a year to answer this question. While seemingly innocuous, a researcher would be unable to use this data set.
e. *Avoid technical terms, acronyms, and jargon:* Language should be easily understood by any participant taking the survey. If technical terms or acronyms are used, a participant may be unable to answer your questions. Do not assume that participants have your level of technical knowledge; instead, use clear terminology. As such, rather than saying:
   - Do patients with arthralgia frequently use the clinic where you are currently working at?

     Try
   - Do patients with joint pain frequently use the clinic where you are currently working at?

   Another alternative would be to take the time and explain the acronym or complex term in your question, rather than assuming the reader already knows what you mean but beware that many participants will just skip over the explanation and try to answer the question as best they can.
f. *Avoid negative questions:* A negative question is worded in such a way that it requires a "no" to respond affirmatively and a "yes" to respond negatively. Questions that are phrased negatively rather than being phrased positively are more difficult to answer and are more likely to be answered incorrectly. As such, rather than saying[3]:
   - Do you believe the government does not do enough to promote healthcare?

     Try
   - Do you believe the government does enough to promote healthcare?
g. *Thoughts on "prefer not to answer:"* Some researchers prefer to compel participants to answer every question and do not include "prefer not to answer" options in their questionnaires. Others believe including

a "prefer not to answer" option increases response rates. Use your judgment: If the question is very sensitive, including a "prefer not to answer," which will ensure that the participants' answer truthfully (rather than answering randomly to get to the next question or dropping out of the survey). If they prefer not to answer, at least the rest of the data you collect will be useful.

h. *Integrate some consistency checks into your questions:* Sometimes, users will answer your questions half-heartedly, more preoccupied with the incentive than the questions they are answering. Others might under-evaluate the commitment needed to answer all the questions and start botching answers: This is especially true if you are running a long web survey. To avoid compiling inconsistent data, integrate some consistency checks into your questions to ensure your participants are answering seriously, and the data you have collected is useful.

   This could include asking the same question twice in the survey and then comparing the responses between both questions to ensure that the participants answered the same way twice. Alternatively, you could incorporate a simple "attention check" question, such as: "If you are reading this question, please select 'Maybe' as your response."

i. *Question placement matters:* The quality of responses can decline sharply for questions that are placed at the end of the questionnaire. As such, it is crucial for important questions to be asked near the beginning of the questionnaire.

j. *Avoid questions that ask too much precision:* While a researcher might really want to have access the most complete data sets, human cognition has inherent limitations in how they can answer questions. Precision in highly complex topics or that have occurred a while back is very difficult to achieve. The following questions illustrate this quite clearly:

   a. *How many times have you opened a PDF since the beginning of the month?*
   b. *How many times do you expect to use our APP in the next 12 months?*

   These questions assume a level of precision that is not possible in an interview. These more quantitative data set can be obtained in other ways (e.g., monitoring usage in a group of beta users), rather than being obtained through traditional market research techniques.

2.3.8 Emoji Facts 😋

Online language has evolved to include emojis. For those unfamiliar with the term, an **emoji** is a small digital picture or symbol used in messages to show feelings, objects, or ideas (such as ❤️). It is increasingly used in communications between individuals: Currently, 92% of the global population utilizes emojis to convey information to users in a way they recognize.[4] They help humanize digital communications.

Since emojis are used to convey emotions, it stands to reason they can be used in web survey to increase participation. After all, 88% of users felt more empathetic toward messages that use these emojis, and 75% feel more connected to messages that use emojis.[5] Consequently, emojis can be useful to illustrate satisfaction scales and can be useful with low-literacy populations. However, moderation is the key. Too many emojis and your web-survey will start looking childish or unprofessional. Also, stick to common well know emojis to avoid confusion.

Overall, emojis offer a free entertaining way for participants to jump into the survey; however, a low barrier to use does not mean to low risk. They can convey multiple meanings with embarrassing consequences, and choosing to use them for the first time might mean you have to be extra careful. Might be best to have someone double-check and ensure that they get the same message you were trying to communicate.

2.4 Incentives

Incentives are an important factor to account for when budgeting for a market research activity. A properly structured incentive can increase participants response rates by 5–25%. Incentives can take the form of money, including monetary compensation, information, specialized services, and more. For example, a 2025 research paper found that the cooperation rate with a prepaid incentive went from 26.6% without incentive all the way to 50.6% with a prepaid incentive.[6] Finally, in voluntary surveys, response rates decline with length, so there is a limit in the information you collect. This is an issue you will not get if you are giving an incentive.

Recent research on the impact of incentives on physician participation in research surveys found that participation rates increased from 1.1%

for physicians offered no compensation up to 2.9% for those offered $75. This both showcased the importance of incentives in increasing doctor participation, while also illustrating the low overall impact of modest financial amount with physicians.[7]

The amount of the incentive also has a significant impact on the response rate. The study above found that participation was at 1.8% when compensated $10, 2.1% when compensated $20, 2.5% when getting $50 all the way to 2.9% for $75.[8] Another study found that offering a more significant gift (e.g., $50 redeemable gift card) was an effective way to motivate clinicians to participate: Out of the 117 clinicians who participated in a survey, 63.5% of participants redeemed the gift card that they received, while the others did not redeem it. The authors believed that this reinforced the need for adequate compensation.[9]

There are several possible incentives you can offer participants, such as:

a. *Direct remuneration:* A monetary incentive can be provided to participants. The Market Research Society[10] advises that any monetary incentive must be "reasonable and proportionate." This implies that incentives should be evaluated on a project-by-project basis, accounting for respondent demographics, subject-matter specialization, and the burden of participation. The minimum compensation for an IDI varies by professional role and interview duration; rates may range from $50 for a non-specialized participant to $450 for a specialist physician. For focus group, the cost per participant range from $100 for standard respondents to $300 for specialized cohorts (e.g., patients with a rare pathology). Professional recruitment may double these costs in certain contexts, representing a significant budgetary consideration.
b. *Vouchers:* As a variation of direct remuneration, vouchers that are redeemable for specific goods are also well appreciated as they are perceived as having monetary value (e.g., a $50 voucher buys $50 of "real stuff"). The advantage for the researcher is that online vouchers are quicker and easier to purchase and distribute (this can all be done online), easier to track and manage (vs. checks or cash), and (sometimes) if bought in bulk, discounts are available.
c. *Non-monetary gifts:* Giving participants a non-monetary gift is especially advantageous in recruitment when there is a clear convergence of interests between the topic, the gift, and the profile of the participants. For example, discussing skincare trends with participants and then giving some of your products. The gift must be

carefully chosen to appeal to the participants (to make sure it is not something the participants probably already have, and to make sure that it does incentivize the participants). Also, there could be some logistical concerns around managing the distribution of products (especially if you must mail them) but there is the advantage of being able to purchase products in bulk.

d. *Free food:* This works best for people participating in a focus group while in a work capacity (e.g., employees) or as part of a community or patient group. You can buy a premium meal than they would usually get for themselves, and it is convivial to "chat over lunch." In these cases, it is a nice touch to add a small participating gift (e.g., a small voucher) as a takeaway for participation.
e. *Charitable donations:* Some individuals with high revenue are unresponsive to standard monetary incentives. As such, offering the budget that you planned for incentives to a charity organization instead might incentivize them to participate. For example, a per diem of $100 might not be enough to encourage doctors to participate in a focus group but donating the $2,000 recruitment budget to a worthy cause and sharing this information during recruitment can sway and increase participation. It appeals to the participant's generous and philanthropic side. It is also possible to offer participants a choice in which charity the funds will be donated to. Once the donation is made, it is important to send a confirmation from the charity (or a contact at the charity itself) to participants so they know it has been conducted.
f. *Sharing the research results:* Some participants might be interested in the results of your findings, as the topic is intersecting with their professional field. As such, it is possible to recruit participants by offering to share the high-level results of the research. In this case, you will have to prepare an executive summary of your findings specifically for those who request it.

Finally, some of the best practices related to incentives include:

1. *Speedy delivery:* Incentives should ideally be distributed within 1 and 2 business days after the data has been collected. Participants should not have to wait for weeks for their incentive to arrive, so do not wait until the end of the project to send incentives: It will be a lot simpler to manage if they are distributed as the interview is completed. It also increases the likelihood of repeat participation in future interviews.

2. *Branded delivery tool:* The email that delivers the incentive should have the organizing company's information clearly identified so that participants can quickly identify the origin, and do not discount the email as spam when they do not recognize the sender.
3. *Properly crafted message:* The message that accompanies the incentive should include a reminder of why they are getting an incentive, and a thank-you note to make them feel appreciated.

The benefits of respecting these best practices include minimizing time spent on customer service for participants inquiring about their incentive, less time spent resending lost or deleted incentives, and increasing the likelihood of participant engagement in future projects. If you are spending too much time managing your incentive programs, investigate some of these tips or into automating some the incentive rewards you use. Personally, I have successfully used Amazon Cards, as they are easy to manage (centralized platform, easy to follow who has claimed them) and they give access to a global marketplace. Other centralized platforms that can be used are Steam or GOG (for younger audiences), or prepaid Visa, Mastercard, and American Express gift cards.

2.5 Data Collection Methods

In this section, we will be going over the basics of many market research data collection tools. These overviews are useful to understand and select the instrument that you need for your project. The methods we will be reviewing are in-depth interviews (IDIs), focus groups, traditional surveys, online surveys, Delphi groups, observation, and mystery shops.

2.5.1 In-Depth Interviews

IDIs are conducted on a one-to-one basis between the researcher and the participant. These interviews consist of mostly open-ended questions. The objective is to explore the topic in a semi-formal format, gathering qualitative information.

The person interviewed could be a potential end-user, a past client, a key opinion leader or anybody else with a relevant relationship to the research topic. Current employees are another good source of information as they are in constant contact with clients and partners. As such, they can provide insight on current customer profiles, goods and services that are popular, satisfaction with current price levels, and experiences with competitors.

While IDIs are more costly (in terms of time and money), they present several advantages over questionnaires and online surveys. For starters, interviews give more opportunities for the researcher to motivate the respondent to participate in a truthful manner, and to not abandon the interview halfway through. Also, interviews allow more flexibility in exploring topics as opportunities in the interview occur. The more exploratory the topic, the more useful the IDI, as it allows the researcher to change the order of questions, to prioritize some topics if time is short, or to go into more depth if the participant is discovered to be someone with an extensive expertise on a specific topic. But the interviewer should be wary of modifying the script too often, since deviating from an existing script too many times can lead to a lack of standardization in data collection and issues in data compilation.

Furthermore, interviews allow more control over the answer process. For example, it might be crucial for a participant to answer a series of questions before another, and an IDI makes it easier for the interviewer to control the flow of responses. Finally, interviews allow the researcher an additional opportunity to evaluate the validity of the answers being given, by evaluating non-verbal cues, or by asking questions again later to validate consistency. This is especially useful when the topic is sensitive, and the researcher expects participants to exaggerate or undermine some facts.

Interviews allow researchers to acquire deeper insights into products or services under development. Some topics that can be explored include:

1. *Purchasing process:* How does the interviewee currently purchase the product? What are the criteria that he looks at the most? Which are the elements that are the most important in the purchasing process? Which things are not important? How important is price in the selection process? Also, when interviewing for the purchasing process, the researcher should remember that there are three types of stakeholder in purchasing decisions: They are champions (who drive the project forward in the organization), influencers (while not making the final decision, are often called upon to give their opinion on a potential purchase), and decision makers (who make the final decision).
2. *Current product:* What does the participant currently use? Why has he chosen to use this product instead of another? Has he tried other products? Why does not he uses these products instead?

3. *Ear to the ground:* What has the participant heard about competitors? Has there been any movement or changes lately?
4. *Innovation:* What is lacking in current products? Is there something the product used to have and does not have any more that irritates the participant? What innovations are coming in existing competing products? Anything that excites him?

2.5.1.1 Keys to Performing a Good Interview

There are several strategic actions a researcher can take to enhance the interview process. Here are a few useful suggestions:

- *Prepare your interview:* Conduct a quick due diligence check on the interview target prior to your interview, to identify potential specialization and fields of specific interest.
- *Record the interview:* When you decide to record the interview, you will be able to focus on the answers the participants are providing, as well as asking questions and exploring topics with your interviewee, rather than spending time taking notes. If possible, use an online platform such as Zoom or Microsoft Teams. These record and offer transcription services, so you can listen to the interviews again later them as needed.
  - *Remember to ask for consent:* Recording an interview without advising a participant is unethical. Furthermore, most people are generally fine about being recorded and appreciate being asked beforehand.
- *Be clear on who and what the study is for, and what the objectives of the interview are:* It is important to share information on what the interview is for, why it is being conducted, and its purpose at the start of the interview. While the knowledge of who is sponsoring the study might polarize some participants, it is essential that the participant be advised upfront. Additionally, it is good practice to remind participants of the specific objective of the interview. Providing both types of information can diminish participant suspicions regarding how the information will be used.
- *Concentrate on the essentials:* Avoid excessively long interviews. A good interview lasts as long as it takes but when preparing your interview guide, plan accordingly. A good rule of thumb is that an interview with an unpaid participant should last 15–30 minutes at most, while a paid participant can expect to spend as much as an hour being interviewed.

- *Listen to your interviewee:* You are trying to collect data, so that means letting the participant share their information. Be careful not to "spoon-feed" the answers that you are looking for. This leads to bad data collection and, ultimately, does not reflect real market conditions.
- *Use silence to get more information:* Sometimes, getting the interviewee to share means not speaking for a few seconds. Research in social interactions has shown that silence is as meaningful in verbal communications as rests are in a musical score. Inserting silences in an interview has a number of benefits: It slows the pace, it is conducive to a more thoughtful mood, it allows the respondent to control the direction of the next steps of the interview, it allows the researcher to note useful associations, and it reinforces the notion of interest on the part of the researcher as they are not perceived as rushing through the interview.
- *Keep neutral:* This is especially difficult when the subject of the interview is being criticized or challenged—particularly if you have a vested interested in the topic—but a successful interview needs to be completed in a neutral fashion. As such, the participant should not be able to guess whether you agree or disagree with their answer.
- *Be encouraging:* This might seem to be in contrast with the advice to keep neutral but the two are not mutually exclusive. Being encouraging means supporting the speaker as they share the information with phrases such as "that is interesting" or "can you share more about that?" Hence, the goal is to keep the interviewee engaged without influencing them by taking a position or by sharing that position.
- *Asking quantitative questions*: For many researchers, asking quantitative questions during an interview can be somewhat uncomfortable. There is always the risk that the interviewee will stop contributing after being asked something they deem overly specific. One way to do this is to ask them for an industry range. As such, rather than asking them, "How many samples does your lab lose to stabilization issues per formulation cycle?", you could ask them, "Can you give me an industry range on how many samples a typical lab loses to stabilization issues per formulation cycle?". By doing this, you depersonalize the question and allow them to answer freely.
- *Stay curious:* In my own experience, many of the most valuable insights have come from following unexpected threads—looking for one thing

and discovering something entirely different. Staying curious, especially during interviews, often leads to information you did not even know you needed.

2.5.1.2 Different Types of Individual Interviews

There are different types of individual interviews, each with their own strengths and weaknesses. We will be touching briefly on face-to-face, phone, and online interviews.

a. *Face-to-face:* Face-to-face interviews are useful when dealing with complex topics. Also, when facing the person you are interviewing directly, you are guaranteed that the person is dedicating all their attention to you. To minimize distractions, you can conduct the interview in a comfortable setting. Other advantages are that this allows the researcher to use visual stimuli (e.g., a mock-up of a potential advertisement or a range of potential packaging) and allows for longer interviews than phone interviews due to the individual proximity. Finally, the personal interaction allows for a more dynamic situation, keeping the participant motivated. The main disadvantage is the cost: It can be costly to travel to each interview (especially if they are geographically dispersed) or to bring every interviewee into your office. Additionally, it can be complex to compile a high number of interviews due to the complexity of the information gathered.
b. *Telephone:* Conducting interviews by phone is both faster and more cost-effective than face-to-face interviews. This method is utilized in situations where multiple people must be contacted rapidly, or when you have a limited budget while still requiring the flexibility of an open-ended questionnaire. The primary challenge is related to attention, as the user can often multitask during your interview. Participants may miss and skip your appointment; as a rule of thumb, one out of every five interviews you book will most likely require rescheduling due to a missing participant (especially if you are not offering an incentive). Finally, it is difficult to pick up on non-verbal cues when performing a telephone interview, so you must be particularly attentive to verbal cues.
c. *Online:* To mitigate the weaknesses of the first two approaches, it is possible to conduct interviews online using software such as Zoom or Microsoft Teams. When performing interviews online with a

camera enabled, the researcher can ensure that he has the complete attention of the participant while also being cost-effective. Also, these technologies allow the researcher to record the interview for future playback and analysis. The downside of online interviews is mostly technology limitations: Quality in audio and video can vary (especially for some locations where the telecommunication infrastructure is relatively weak), Consequently, you may encounter intermittent errors that limit the information you get from visual cues or lead you to miss some comments due to audio problems. However, with the advent of AI-enhanced transcription and summaries, these platforms are rapidly becoming a market researcher's best friend.

2.5.1.3 Motivating Interview Respondents

Some interviews will be more difficult than others. Some individuals will have changed their minds about participating; some might do it solely for the incentive, while others might be compelled by a superior or colleague into participating. Whatever the reason for the resistance, it could make the interview difficult and quite uncomfortable. Here are some tips that might help you motivate the respondent and ensure that you get the information you are looking for.

- *Increase the participants' interest:* A participant might have low interest because he fails to see the relevance of the answers he is sharing or has been compelled to participate. Supplying context on why the interview is occurring and mentioning the connection between the individual and the project can motivate the respondent. For example, reminding the participant that the objective of the project is to develop a technology that will simplify the participant's daily work could help motivate him.
- *Recognizing the participant's expertise:* Reminding the participant that their specific expertise was the reason they were selected for the interview can enhance their sense of professional recognition. This can be conducted not only at the start of the interview but throughout the interview as well.
- *Reducing threats to the ego:* If you feel the participant's ego was bruised by a previous question, it is an effective technique to include a "face-saving" preface to mitigate the impact. For example, you could start a new series of questions with a statement such as: "As you know, topic ABC is not a subject many researchers are familiar with. However, I

believe you have significant insight into it, so I am looking forward to hearing your thoughts and experiences."

- *Reducing the issues related to perceived authority:* There may be situations where the participant identifies the interviewer with a specific role or position that influences the interview process. For example, the interviewer could be a key opinion leader or a former teacher collecting data from past students. Clarifying the interviewers' role in the interview process throughout the interview can enhance the interview flow.

2.5.1.4 Recruiting for Interview

Selecting the right participants is a critical step in ensuring the relevance and quality of insights gathered during market research or technology evaluation. Ideally, participants should be individuals who will actually use the technology once it is deployed. When the project is still in an exploratory phase and the end-user profile has not yet been fully defined, the focus should be placed on individuals who are considered strong potential users or influencers.

A well-balanced mix of participant profiles is essential. The group should include end users, key decision-makers, internal champions, and technical experts when they are directly involved in decision-making or implementation. This diversity helps capture a more complete view of the user journey, the decision process, and perceived benefits or challenges.

It is also important to include both novice and experienced users. Novices can help identify challenges related to initial usability or onboarding, whereas more advanced users can highlight higher-level improvements and benchmark the solution against existing alternatives.

Finally, common selection biases must be avoided. Recruitment should not be limited to internal "supporters" or individuals who are already predisposed to view the solution positively. Likewise, an over-representation of highly technical participants can distort results when the goal is to assess usability, comprehension, or practical value. A careful participant selection process, aligned with the study's objectives, is ultimately essential to generate valid insights and support informed decision-making.

2.5.1.5 Interviews in Life Sciences

IDIs in life sciences are a popular way to get the information you need, especially when dealing with topics that are sensitive in nature. It is a great way to speak to people confidentially and get their views on healthcare topics, such as their personal health and that of loved ones, their use of pharmaceuticals, and so on.

Interviews are also useful to discuss topics for which an "inch-deep, mile-wide" approach is needed. Sometimes, a market researcher will have little information on the topic he is researching and asking a few general questions will enable him to quickly identify opportunities.

Interviewing doctors and medical personnel can be especially challenging as these individuals are constantly solicited for their time. As such, you might have to set aside a an important per diem to interview them, ranging anywhere from $150 to $400 per interview (and even more for key opinion leaders). Here are some things to keep in mind:

- Identifying doctors to interview can be challenging: One way to save time is to purchase a contact list from a specialized provider. You can customize your purchase to your specific needs such as by medical specialty or geographic region). Be sure to double-check how often the data is updated, as well as what contact information they will be supplying (email, physical address, phone, fax number, etc.). Additionally, verify if the doctors have opted in to be contacted by third parties or if the information was gathered from the web at large (i.e., scraping). LinkedIn is also an effective platform to identify and connect with participants.
- If you want to build your own list, professional associations, licensing boards, or lists of attendees at a conference are good starting points. Reviewing scientific publications through a database such as Google Scholar, PubMed, or Europe PMC can also be useful to identify key opinion leaders and topic experts, as well as for obtaining their contact information.
- The best way to reach doctors is by email. Phone calls often end up in voicemail or are intercepted by gatekeepers (e.g., administrative assistants).

A typical email might look like the following:

> *I am conducting a study on behalf of Bio Biotech regarding the technologies currently used to diagnose XYZ. The results of this study will be used to help our client develop a new diagnostic device to better meet your and your patients' needs. We value your input and would like to arrange a 20-minute telephone call to discuss the topic*

with you. Please note that all results will be anonymized, and only consolidated results will be shared with the client. In recognition of the value of your time, we are offering an honorarium of x$.

Another way to recruit specialized participants is to use one of the many online platforms where experts are available for consultations and single interviews. The platforms, such as Zintro (www.zintro.com) enable you to easily post projects and contact subject-matter experts directly, acting as an escrow service. You can find scientists, doctors, and more on these platforms; experts can either respond to your inquiry or refer it to other qualified professionals within their network.

Finally, do not get discouraged if you get low traction for your solicitation. Remember that doctors are a heavily solicited population and that it will require a significant recruitment effort to obtain your target sample.

2.5.2 Focus Groups

A focus group is a small group of individuals brought together to discuss a specific topic. The added value of a focus group (versus individual interviews) is that the interaction between individuals generates a wealth of information for the researcher. As such, focus groups are useful to gain a better understanding of what people are thinking and why they are thinking it. This format is also beneficial as many participants will generate more information in a group setting where they feel safe and they do not feel that they are the sole focus of the interview. Finally, the information gathered in a focus group can be very useful when you are design a quantitative questionnaire or interpreting the information gathered from a quantitative research project.

The main issue with focus groups is similar to the challenges found in peer groups. Factors such as social pressure (the desire to conform to the group and avoid disagreement), an individual's domination of a group, or the "halo effect" generated by key opinion leaders can all impact the quality of the interactions and the information generated. This phenomenon is also called groupthink, in which members are more interested in getting the approval of other group members rather than trying to share their true thoughts. This is especially apparent in groups where a senior key opinion leader is inserted in a group with more junior equivalents; in essence, people think if they do not conform, they run the risk of appearing.

It is the role of the moderator to step in, rebalance the focus group, and ensure that it does not become biased. The other issue to remember is that while focus groups can be used to evaluate a group's feelings or views on a topic, it cannot be used as a final tool. It is exploratory in nature, not statistically valid, and the information you gather, while invaluable in interpreting existing data or setting up more research, cannot be an endgame.

Ideally, a focus group should consist of 8–10 participants, as the group needs to be sufficiently large to generate dynamic conversations but not too large as to leave some participants out or to become difficult for the moderator to manage (as parallel discussions start to be generated). Smaller groups of five to six participants are also possible but create a different group dynamic; these are more useful when focusing on a highly specific topic or if they include a product-testing component.

About 45–75 minutes is the ideal time frame for a focus group: too short, and you will not generate any deep insight, while too long, you risk participant fatigue, where they quickly agree in the hopes that the focus group will end. Three to four focus groups are usually necessary to fully explore a topic: Beyond this point, you may find the same information being repeated.

In life sciences, focus groups are especially useful for interacting with end users—such as doctors, nurses, laboratory technicians—as well as for gathering feedback from patients. As such, they can generate valuable information related to marketing, branding, competitors, and product issues.

Consistent with Kelly & Rupert's (2009)[11] observation that physicians present themselves as rational decision-makers in focus-group environments, Kerasidou & Horn (2016) showed that physicians emphasize rationality and emotional control when interacting with colleagues[12]. As such, focus groups are not an ideal environment for eliciting emotional information from doctors. Also, it can be especially challenging for you to reach and recruit specific participants.

2.5.2.1 Guidelines to Preparing a Focus Group

A focus group discussion centers on a predetermined set number of questions: anywhere from 8 to 10 questions is a good number of questions. It is important not to overload the discussion guide with too many questions, as the moderator might not have sufficient time to address every topic or might feel rushed to touch on every topic, potentially missing key insights.

When writing questions, keep them open-ended: Participants should not be able to answer them with a simple "yes or no." Also, participants usually do not have a visual copy of the question they are discussing, so try to keep them simply phrased. If necessary, ask follow-up questions to address complex topics. A reliable methodology to use when building your discussion guide is to have three groups of questions:

1. *Engagement questions:* Simple questions to warm up the group, so they get to know each other and the topic of discussion, making them comfortable.
2. *Exploration questions:* Main questions for the focus group on the topic the researcher is researching.
3. *Exit question:* The last question to see if the participants want to add anything relevant to the topic before ending the focus group.

Table 2.3 provides a sample discussion guide for a focus group exploring patient engagement with store-brand headache medication.

2.5.2.2 Recruiting for a Focus Group

The participants' profiles as well as inclusion and exclusion criteria should be decided before the recruitment process begins. This information will be

Table 2.3 Example of a Focus Group Discussion Guide Exploring Patient Engagement with Store-Brand Headache Medication

Engagement Questions • What is your favorite brand of headache medication? • Which store do you usually buy headache medication from?
Exploration Questions • How do you choose which medication you will buy at the store? • Has anybody here ever tried store-brand headache medication? What was your experience with it? • What are your thoughts on store-brand headache medication? • How safe are store-brand medications compared with brand-name medications? • What are the pros and cons of store-brand medication?
Exit Question • Is there anything else you would like to say about headache medication?

useful to screen and select participants. Examples of criteria include the sex of the participant, age, user versus non-user, or any other focus you have defined earlier in your market research objectives. There are many ways for you to recruit participants for a focus group; some examples are included here for your convenience:

a. *Nomination:* Participating key opinion leaders can nominate other applicants they believe would make good contributors. Nominated individuals are interesting potential panel members as they are expected to already have a deep knowledge of the topic at hand: they are also more likely to participate since they have been referred by a noted expert. However, this has the unfortunate side effect of potentially creating bias in the sample, as individuals are more likely to nominate individuals who agree with them rather than identify participants who are openly disagreeing.
b. *Random selection:* If you are working with an especially large pool of willing participants, the researcher can select at random participants out of a preexisting pool (e.g., employees in a company) until the right number of participants is attained.
c. *Members of a group:* In some cases, an existing group of participants can be a great recruitment pool from which to invite participants. For example, a non-profit association or a chamber of commerce can both provide interesting pools of participants.
d. *Volunteers:* When you are conducting a broad research effort, participants can be recruited at large through traditional means such as—flyers, newspaper advertisements—as well as electronic means, such as Craigslist, online job boards, and specialized discussion groups.

When forming your groups, it is good practice for you to over-allocate and include one or two more participants than needed: It is not uncommon for 20% or more of a group does not show up (even when they are incentivized), so overbooking slightly mitigates this.

Additionally, do not forget the multiple costs associated with running a focus group, such as recruitment costs anywhere between $250 and $750 per panel if you are dealing with a specialized third party), recruitment and incentive fees, facility costs (room, food, recording), direct moderator costs (which can range from $750 to $1,500 per session for skilled moderators), and indirect costs (transcription, analysis, and final report). Hence, running a series of focus groups can be a very costly endeavor particularly for startups.

The two variants discussed later—online focus groups and triads—are interesting alternatives if you are unable to shoulder the full costs associated with traditional focus groups.

2.5.2.3 Running the Focus Group

Ideally, a focus group is moderated by two persons: The head moderator and an assistant moderator:

- The *head moderator's* responsibility is to facilitate the discussion and to directly intervene with the group. He listens and ensures that all participants participate. He should have solid knowledge of the topic to be able to ask follow-up questions. Skilled moderators will be able to paraphrase long and complex comments and build on these to move the conversation forward. Finally, as the moderator is in a position of authority (he is moderating and handling traffic in the conversation), he must always stay neutral and refrain from agreeing or disagreeing with the comments as they are shared.
- The *assistant moderator* takes notes and handles the recording (if the focus group is being recorded). He listens, notes, and observes any subtle events that might occur. It is preferable that he lets the moderator talk and handle the conversation but must be ready to add his input if needed.

For analysis purposes, it is useful to collect participant's demographic information. A short questionnaire with a few questions is all that is needed. The assistant moderator must make sure that participants have answered the questionnaire before coming into the focus group session or has them do it on-site before the focus group begins (Table 2.4).

It is also necessary to have the participants fill out a consent form. The consent form informs participants that they have a responsibility to keep all discussions private as well as an agreement not to discuss any information discussed during the focus group outside the focus group setting. It can also include an authorization request to quote participants (anonymously or directly).

Once consent forms and demographic sheets are collected and reviewed, the focus group starts. The moderator welcomes participants, reminds the participants of the focus group objectives, and sets some ground rules. These ground rules could include:

Table 2.4 Example of a Demographic Form for a Focus Group with Physicians

Physician Focus Group Participant Demographics	
Date:	
Time:	
Location:	
Your gender	Your age
• Male • Female	• 30–40 • 41–50 • 51–60 • Over 60
Type of practice • Public • Private • Other: ____________	What is your specialty? • Cardiologist • General physician • Other: ____________
How long have you been practicing? • Less than 5 years • 5–10 years • Over 10 years	How many CME credits did you complete last year? • Fewer than 25 credits • 26–50 credits • Over 50 credits

- I would like everyone to participate: Please do not interrupt another participant while they are sharing their opinion.
- There is no right or wrong answer: Everyone is sharing valuable insights and information, so please refrain from judging anyone's opinion.
  - You are free to disagree! Whether you agree or disagree, please share it with us—we want to know! Tell us why, in a non-judgmental way.
- This meeting is confidential. What you share with us is confidential, and when we share it with others, it will be conducted in an anonymous manner.
- We will be recording the conversation: We want to be sure we do not miss anything. But do not worry, there are no names in the final report, and everything will remain anonymous.

Immediately after all participants leave, the head moderator and assistant moderator should take a moment to debrief while the recorder is still running, while their reaction is still fresh; this debriefing will be an invaluable part of the data collected during the focus group.

2.5.2.4 Variation #1: Online Focus Groups

Online focus groups are a variation of traditional in-person focus groups. As their name implies, online focus group participants engage online using a webcam and microphone. Today, most groups will use a common web conferencing tool such as Zoom, Google Meet, or Microsoft Teams.

The advantages are numerous. First, there are lower costs associated with online focus groups; some estimates are that online focus groups cost about half as much as a traditional focus group. Additionally, online focus group recruitment is faster and the time commitment is less intense. This is especially useful with professionals (e.g., doctors) who have limited time to participate in focus groups and are historically difficult to reach. Similarly, individuals with specific characteristics (such as those with rare diseases) may not be in close geographic proximity to one other. Online focus groups make it easier for you to bring these individuals together and make it easier to reach a representative sample size. Moreover, people who are more reserved may be more willing to speak in an online setting, as they often feel less directly judged by others (not being in the same room and not being in direct contact). Finally, observers, clients, and the assistant moderator can monitor the session on a separate feed and provide comments to the moderator discreetly. Note that if the client is observing a direct feed, you must provide notice to the participants.

Nonetheless, there are several challenges associated with online focus groups. For starters, as participants are not in physical proximity to one another, there is a limited opportunity for direct interaction. As such, a lot of the direct interaction (which is sometimes the richness of a focus group) is lost in an online setting. Critics of online focus groups will go as far as claiming that the interaction in an online focus group is insignificant. Additionally, populations who are technologically challenged may be unable to participate effectively in online environments. Therefore, if you are seeking feedback from an older population, online focus groups might not be an effective solution for you. Furthermore, technology limitations such as—poor internet connections or low-quality webcams and microphones—can all have an impact on the quality of the focus group session and can critically impact the interactions, sometimes even breaking the "flow" of the conversations. Finally, it is particularly difficult to act on non-verbal cues in an online setting.

To ensure the best results when you conduct an online focus group, several guidelines need to be respected:

1. Groups must typically contain between six and eight individuals; however, the researcher might need to over-recruit one or two participants to account for no-shows, as these are more frequent than in traditional focus groups. During the recruitment process, the researcher should ensure that the participants have the minimum technology requirements and knowledge to be able to participate.
2. During scheduling, the researcher should account for time zone differences between participants when selecting the time slot.
3. It is essential that you test all equipment and software before the actual online session. For testing purposes, the researcher can recruit a few of the scheduled participants to participate in a testing session to ensure that the moderator, clients, and other stakeholders have an adequate handle on the technology.
4. Finally, the researcher should ensure that he sufficiently communicates with participants. This includes sending reminder emails and detailed explanations on how to log in to the system.

During the presentation, the moderator can use PowerPoint slides to share visual content, and he must be sure to keep an eye on the participants in attendance. It is his responsibility—or the assistant moderator, if there is one—to ensure that participants are attentive and that they are not multitasking, such as checking their email or playing games).

2.5.2.5 Variation #2: Triads

Triad focus groups consist of three participants and are utilized in very specific situations.[13] Because there are fewer participants, these sessions can be shorter—getting everything you need conducted in under an hour—allowing for more groups to be seen in the same time frame. Additionally, the presence of fewer participants allows the researcher to perform more in-depth research and probing, while limiting groupthink (participants retain more of their individual position, and outliers are less influenced by the general agreement of a large group). Finally, triad focus groups allow more dedicated product testing and observational usage; as each participant has more individual time to test the product, rather than having a few participants test the product and the other participants observing it.

The main disadvantages are that since the groups are smaller, more of them need to be conducted to obtain sufficient volume of information. Triads should not be used as a cost-saving measure but rather as a distinct methodology to obtain the same information. As a rule of thumb, you should hold enough triad focus groups to equal the participant volume of one traditional focus group (e.g., three triad focus groups for every one regular focus group). Furthermore, triad focus groups can be more "awkward" to manage, since there are fewer individuals, putting more pressure on the moderator's skills to keep the discussion moving forward.

2.5.3 Traditional Surveys

Traditional surveys are a legacy tool used to collect a large amount of quantitative data from a target audience. Using a questionnaire, the researcher develops questions that address the topic as needed: Most questions should be closed-ended to allow the participants to complete the survey rapidly and to allow for quick data compilation but there is the possibility of having some open-ended questions as well. The two main advantages of traditional surveys are that they are economical and that they can be conducted in a manner that protect the identity of respondents.

Questionnaires can be prepared and distributed much more economically than one-to-one interviews. For example, if a researcher is administering a survey by mail or soliciting a large group of individuals at a conference, the lower costs are distributed over a larger sample of participants. Secondly, questionnaires can be sent to potential participants with a pre-addressed envelope, which participants can send back to the market researcher without any identifying features.

However, the use of this tool is increasingly niche, and in my experience, almost never used. The major reasons you would conduct a direct-mail survey are: (a) Your audience has low or uneven internet access (e.g., older adults, rural, lower-income segments). (b) Your sampling must be tied to a physical address, such as residents around a hospital or clinic. (c) You are leveraging a brand where legitimacy matters for the survey (ex: the survey is for a government agency or recognized company and physical materials feel more official and credible).

There are several different techniques that can be used to distribute and collect information via questionnaires. We will touch briefly on those used less often in life science, and we will discuss the most useful method—online surveys—in depth as a standalone section.

2.5.3.1 Direct Mail Surveys

Direct mail surveys are provided physically and directly to participants, either by handing them out in person or sending them through the mail. They have long been a technique of choice because they are less expensive, have a wide geographic reach, and eliminate the risk of interviewer bias. Nonetheless, their current effectiveness is questionable; generic direct mail surveys usually have a 2–4% response rate. Furthermore, there is a lengthy response cycle between the survey being first sent out and getting the response back to the market researcher, and you often need to engage in constant reminders—via phone or email—to increase participation. In a survey of targeted patients, researchers found that the response rate was 24.3% for patients, 6% less than those responding through the web (30.6%).[14]

While they have been largely replaced by online surveys but could be useful in the specific situations discussed previously. To increase participation, a Media Logic study found that envelopes that grabbed attention, seemed friendlier, and stood out more were more likely to be answered by seniors in a healthcare setting. Additionally, letters that look easier to read, are quicker and easier to understand, and have a better layout were more appreciated.[15] Another way to increase participation is to include a pre-paid envelope as well as personalizing the letter that accompanies the questionnaire.

2.5.3.2 Telephone Surveys

Telephone surveys used to be a cost-effective method for collecting data, particularly when trying to quickly reach a population at large. They allowed for the rapid collection of data, could be used to address complex topics, and enable specific targeting of participants.

Nonetheless, there are fewer landlines, making populations difficult to reach. Furthermore, participants are more likely to drop out halfway through or provide inaccurate responses, especially if the phone survey was longer than initially announced. Finally, participants are less likely to answer if they do not have a relationship with the caller (i.e., they do not recognize the caller ID) and timing is everything when conducting a phone survey: Calling participants at an inappropriate time will most likely result in the caller refusing to participate. The emergence of do-not-call list and rise in fraudulent calls have contributed to reducing the efficiency of this tool.

However, companies and organizations that have pre-screened lists—such as members, clients, or opted-in individuals—can still have successful telephone survey campaigns compared to those conducting cold calls. Another way to increase participation rates, you should prepare lists of questions that are simple to answer and keep the survey short. Remember that if you are speaking to people representing organizations, you must make a specific note of which individual you interviewed.

2.5.3.3 In-Person/Street Surveys

In-person surveys bear a resemblance to IDIs but focus on quantitative data. They consist of approaching people in the street or in a natural setting and asking them for a moment so you can ask them a few questions. This is the main strength of in-person surveys: They can be conducted anywhere. For example, if you want to open a healthcare facility or sell a product locally, it could be a good way to validate market demand by visiting the chosen location and surveying people around the location to get a sense of the "walk-in" traffic potential—number of potential clients who will enter a facility as they pass by it. This method is also useful to quickly collect data on a generic topic. It allows for the use of follow-up questions adjusting the questions in a fluid manner. It also permits the use of simulated materials and mock-ups as the participants are facing the interviewer.

However, it can be difficult and frustrating to recruit participants. Some things you can do to attenuate this include carrying an official ID, surveying the site in advance, and keeping the interview under 10 minutes. A small (visible) token of appreciation can also enhance participation. Another challenge is that in-person interviews can take much longer to complete than a web survey. It is also more costly, especially when you include data compilation and the interviews you will have to reject due to invalidity concerns. Nonetheless, using an in-person survey allows some innovative practices—such as starting a survey in person and asking participants to complete it online later, or the use of mock-ups—and can be useful in very specific situations.

2.5.4 Online Surveys

The rapid pace and development of technology have created new opportunities for collecting data. As such, the use of online surveys has grown immensely in popularity. They are cost-effective, simple to use, and—if conducted properly—can reach a wide range of populations. They allow your participants to complete the survey on their own time with minimal effort. Additionally, they allow for the use of various media (e.g., audio and video), making the experience richer for the participant and allowing more complex data gathering for the researcher. Most online surveys today are conducted using web-based survey tools, although the popularity of surveys on mobile platforms are rising. This has the advantage of allowing for participation tracking, enabling you to directly target non-respondents, send them reminders, and also allowing for direct data entry, making them much more efficient.

However, there are limitations to using an online survey for data collection. For starters, they are much more effective when used with a closed population (e.g., employees, members, or clients), where the response rates tend to be higher. Also, online surveys are focused on populations that have access to web-based technologies: Hence, online surveys will usually skew toward populations that are younger than 65 years old, college educated, and have higher-than-average household incomes. If you are not targeting one of these groups, another data collection tool may be more appropriate. Finally, with the rise in popularity of online surveys, there is a noticeable trend toward over-solicitation of users, leading to user fatigue: Proper incentives (as discussed in Section 2.4) are becoming a key factor in increasing user participation.

There are considerable challenges in recruiting participants for life science surveys, particularly clinicians. A study conducted in 2015 found that an online web survey targeting clinicians yielded a 35% participation rate, with deep variances across specialties: Ranging from 46.6% (neurology/neurosurgery) and 42.9% (internal medicine), 29.6% (general surgery); 29.2% (pediatrics), and 27.1% (psychiatry). Lack of time and survey burden were the most common reasons for not participating.[16]

Another study found that general practitioner survey rates could be increased with incentives (larger and upfront, if possible), peers pre-contacting targets via telephone, personalized packages, and sending the survey on a Friday.[17] A third found that survey that used personalized invitations had higher response rates than generic links (40.2% versus 25.3%).[18]

2.5.4.1 Tips for Designing Your Web Survey and Increasing Your Response Rates

There are several key strategies to remember when building a web survey to increase your response rate:

1. *Keep the survey short and simple:* This is the most effective way for you to reduce user attrition. While you may be able to utilize a longer survey if you are providing an incentive keep in mind that the value of the incentive is directly related to the length of the survey.
2. *Work on your formatting:* When designing your survey, ensure it is visually appealing to participants and easy to read: The easier it is for the participant to read and navigate, the less attrition you will get throughout the survey.
3. *Be straightforward about the time to answer:* Announce upfront the length of the survey and, if possible, use a progress bar on top of the survey to keep participants engaged.
4. *Optimize your survey for mobile devices:* Ensure you use a survey platform that will optimize your survey for mobile platforms: More people use mobile devices to complete mundane tasks while waiting in line or commuting, for example, and if your survey does not display properly on a mobile device, they might simply drop out and move on.
5. *Include an end of survey message:* The end of survey message should include a sincere thank-you, a reminder of the objective of the survey, and your coordinates or a link to a dedicated email address (e.g., survey@yourcompanyname.com) so participants who have concerns or questions can easily contact you.
6. *Send reminder mails:* Send the first reminder email 1 week after the initial invitation, followed by a final reminder 1 week after that. You should expect diminishing returns of approximately half each time you send out a reminder. So, if the first email got you a 10% response rate, the first reminder should get you an additional 5% of respondents, and the last email another 2–3%.
7. *Plan the first survey invitation:* When sending out web surveys, target Mondays for B2B web surveys, as this will increase your response rate. If you are targeting end-users, Wednesdays and Fridays are preferable to get the best response rates.
8. *Send a personalized message:* As discussed earlier, a personalized message could increase participation by almost 15%. Hence, directly

inviting and reminding participants to answer your survey will likely result in a substantial increase in the response rate. Related to this, it appears that refraining from relying on intermediaries to disseminate survey links could also increase participation.

Finally, you must ensure that you have informed consent and comply with anti-spam regulations. Informed consent is achieved by clearly explaining the survey objectives at the start of the survey, providing the identity of the sponsor of the survey, explaining how the data will be used as well as reiterating the confidentiality agreement surrounding participation in the survey. You can include a "do not agree" box at the start of the survey to discard participants who do not consent to the survey conditions. Regarding anti-spam regulations, these vary by jurisdiction to the next, and only apply if you are surveying a population that has not opted in.

2.5.4.2 Mobile Online Surveys

Mobile online surveys utilize mobile technologies—such as smartphones and tablets—to collect data. The growth in popularity of these devices has created new opportunities to reach participants and collect data rapidly: Pew Global reports that an overwhelming majority of the population in almost every nation they surveyed owns some form of mobile device, even if it is not considered "a smartphone."[19]

Mobile online surveys present multiple advantages. They can be used to reach populations that are increasingly hard to reach with traditional market research methods (i.e., younger audiences). Also, they usually have higher response rates as mobile surveys are convenient to answer. They also have lower overall costs as it takes less effort to reach the minimum sample size, and mobile surveys can use some of the device's features (e.g., the global positioning system (GPS), the camera, or the microphone). Finally, as participants often have their mobile device on them, they respond quickly, providing you with immediate feedback and even enabling event-based surveys.

Some of the downsides of mobile marketing research are reduced response rates depending on the group targeted as some demographic groups are underrepresented on mobile devices. The questionnaire must also be shorter and simpler, which can make it difficult to collect in-depth data. Furthermore, some practitioners have found that higher incentives are necessary to gain sufficient traction. In addition, some people find receiving online survey invitations particularly intrusive (akin to receiving a phone call during dinner). The overuse of mobile online surveys may lead to members or customers opting out of getting contacted in future communications.

There are a variety of ways to collect participants for the survey, ranging from push methods (where the survey is sent out to participants independently of location and actions) to pull methods (the survey is sent to participants in a specific location, or those purchasing a specific product).

2.5.4.3 Getting Participants

Once your survey is built, the next step is to get participants to answer your survey. If you are surveying existing clients or interested parties, you can invite the participants directly to participate in your survey. However, if you need to survey general population, there are several professional solutions that are available to you.

The first solution is to purchase a list or database of pre-qualified email addresses. Numerous private organizations sell these lists of pre-qualified participants and databases that can be used for survey purposes. The price of these lists can vary a lot, from a $100 to $1,000. You must remember that purchasing a list does not automatically grant you authorization to contact those participants, so it is important to verify whether the database consists of participants who have opted in to receive communications from third parties.

Furthermore, if you are sending out your survey to a population at large without any consent of solicitation, you may run afoul of anti-spam regulations. Make sure you comply with existing regulations by adhering to local requirements, such as including a correct header and a functional unsubscribe link. Generally, if you are contacting individuals with whom you have an existing relationship (e.g., customers, members, or employees), they are usually excluded from such regulations. Double-check your relevant laws to be on the safe side if you decide to utilize an "at-large" data collection strategy.

The other alternative is to use a third party to recruit participants. For example, SurveyMonkey offers this service, with costs typically ranging from $1 to $7 per participant, depending on the criteria of the participants you need. Another alternative is Pollfish, which utilizes proprietary and third-party mobile phone applications to recruit and validate participants. They currently have a database of millions of users throughout the world. In both cases, expect anywhere between 24 hours and 2 weeks to reach your target sample size. Because you only pay per completed survey, not per respondent (e.g., you do not pay for incomplete surveys or those who did not pass the screening qualification), these recruitment services are quite affordable for startups trying to get a better perspective on their market. They are quite efficient if your end client is the general public but more specialized populations could be a challenge, be a little bit more expensive or take more time.

As an example, in a project from a few years ago, a client needed to better understand the purchase process of skincare products for patients undergoing radiotherapy. The secondary information available on the topic was quite limited, so we engaged in a web survey online using a third-party recruitment service. The simple web survey was online within 24 hours; within 1 day, we had a sample of 150 people sharing insights on their purchasing process, favored brands, and pricing sensitivity for less than $300. In another instance, I needed to research use of health supplements in Tunisia. Using one of these platforms allowed me to quickly and effectively gain access to hundreds of Tunisians using health supplements. Within 48 hours, I had a survey with 400 respondents sharing their insights.

This type of personalized market research would have been impossible just a few years ago and speaks volumes about how emerging technologies are shaping the way we do market research.

2.5.5 Delphi Method

The Delphi method is an interesting methodology that is used to get information from experts and key opinion leaders regarding future trends. It is particularly suitable when the researcher is faced with limited secondary information or an exceptionally complex topic.

The methodology is simple. A group of experts are individually asked a series of open-ended questions on a specific challenge (usually by email, although this was traditionally conducted via mail). The moderator then compiles the answers he has obtained, identifying the majority position of the group, listing the arguments for and against the positions. He then shares the majority position with participants as well as generating a new round of questions, which he sends out to the participants for additional feedback. As such, the participants are confronted indirectly and face the opinions of other experts in an anonymous setting. This process is usually repeated until a consensus forms around the topic being discussed. However, some researchers suggest that a maximum of four iterations should be conducted before concluding a Delphi group. Also, divergent opinions should be included in the report with their supporting arguments.

Hence, a Delphi group involves the systematic questioning of renowned experts to explore a breakthrough topic. This methodology is particularly useful for forecasting trends and engaging experts who are difficult to reach individually, as responses are completed at the convenience of the participant while conferring anonymity on the participants. Typically, participants are not revealed to each other, even at the end of the project. The systemization of data collection, combined with this anonymity given to participants minimizes the halo effect that some key opinion leaders can have on other participants and enables participants to re-adjust their opinions more freely throughout the process. It is also relatively inexpensive.

However, the Delphi method results are tied to the competency of the moderator managing the group. An inexperienced moderator might lead the group toward a false consensus or lose participants as the group progresses. Also, there is a natural tendency to eliminate more radical options and focus on more mainstream compromises. Finally, the method is more time-consuming and lengthier than a traditional focus group.

In this method, the bulk of the work is conducted by the moderator. He is responsible for generating the initial questionnaire, following up with participants, compiling the information as it is generated, and animating the panel. He is also responsible for coaching participants and training them if they are not familiar with this methodology. His skills in written communication will have a direct impact on the success of the panel.

The researcher should be careful when recruiting participants and he should make sure they have been informed that the process is time-consuming. He should account for participant attrition, and plan accordingly by initially over-recruiting. While some researchers advocate for the recruitment of up to 30 participants initially, too many participants lead to complexity issues in data compilation and coordination: up to 15 participants is a good starting point, which considers participant attrition.

One final thing to note is that the Delphi method is highly recommended for use in a healthcare setting as a reliable means of determining a consensus for a defined clinical problem. However, it may have limited practicality in market research situations where you are looking to generate some market-oriented data (e.g., market size and growth).

2.5.6 Observation

Collecting data through observation enables the researcher to monitor and see how a product is used or a situation is handled in the real world. While the researcher is watching how individuals react in everyday occurrences, he gains valuable insight. One of the main strengths of observation is that the data being collected does not depend on the impressions or knowledge of the individuals being observed; rather, it depends on the interpretation of the researcher of the action they are performing. Hence, the researcher is not contingent on what the participant has done in the past or claims to have done, does not have to deal with credibility issues, and memory lapses do not influence the quality of the information gathered.

There are several parameters that the researcher controls when starting an observation activity. He can declare the observation activity to the participants (the participants are made aware that they are being observed) or he can do it discreetly (the participants do not know they are being observed). The advantage of hiding his observation activity is that he will not influence and modify the normal course of the activity but the downside is that he loses the opportunity to ask questions as events occur. He can also do observation in a natural setting (the participants are in their normal environment) or in a laboratory setting (the market researcher sets up a scenario and then observes the actions of the participants). Finally, the researcher can choose to directly participate in the activity to increase his comprehension of the phenomena. Participation observation allows interaction while recorder observation is easier to analyze. For example, a market researcher could work as a volunteer in a hospital to better understand the patient waiting experience by chatting with patients who have been waiting for some time. In this case, the observer must be careful not to "contaminate" the observation experience or influence responses and must be mindful of potential bias: It is quite easy to become attached to individuals, which can bias the evaluation process. Finally, the presence of the participant–observer can change normal dysfunctional patterns or functions, and the researcher should account for the impact that his presence will have on normal processes.

Coming back to the earlier example on advanced insulation, my client was developing a very robust insulation container and wanted to see if it could be used by fishermen. To understand how fishermen use existing insulation containers in their everyday setting, I visited a dock and observed fishermen coming back from fishing trips. The objective was to record and observe how the fishermen handled their insulation containers after a day at sea. The observations were surprising: In some situations, the fishermen would simply throw their containers overboard when arriving at the docks from the sea. I also noticed that quite a few containers were pierced by the forklift carriers. As such, while in literature, competitors claimed that their products have a useful life of 10 years, our observation at the dock revealed that everyday usage was brutal and incompatible with the product my client was developing.

One key difference between my client's product and traditional insulation is that a normal insulated container can be pierced by a forklift, "patched" with insulation foam and some duct tape, and then sent back to the docks. It would still have an estimated 75–80% efficiency rating, whereas my client's technology required 100% product integrity. The industry's definition of functional included heavily damaged containers—a fact that observation brought to light (Figure 2.1).

There are some conditions to using the observation method. When choosing this methodology, the researcher should ensure that the information is observable, or inferable: Choosing to observe a customer's purchasing decision process would not be possible but observing how they use a product is possible. Also, the phenomena to observe must be frequent, repetitive, or predictable. If not, it becomes too costly to observe. Finally, the observation must be relatively short in duration to be cost-effective.

As mentioned earlier, observation has a number of strengths over other tools. First, it allows direct access to the person being observed without the use of intermediaries. As the researcher is observing, there are fewer

Figure 2.1 A few containers damaged but still used in a day-to-day situation by fishermen.

opportunities for the subject to respond dishonestly. Finally, it allows the researcher to observe the subject in his normal environment, which allows him to notice environmental factors or behavioral nuances.

The disadvantages are also steep. First, as the observer is collecting the information, he is inserting himself as a part of the data collection equation. His point of view, past experiences, and even mental state—such as fatigue, emotions stress—can all taint the information he is collecting. A researcher observing a setting that he has himself experienced in the past can have preconceived expectations that he specifically looks for. In addition, the presence of the observer has a direct impact on the subjects he is observing, especially if they are aware of his presence and are constantly questioned about decisions they make or actions they take. Finally, there is an undeniable cost to performing observation, both financially and timewise. Collecting significant observational data can take a lot of time on the part of the researcher.

2.5.6.1 Use in Life Sciences

Observation can be used in a variety of situations. For example, it can be used in environments with patients, observing how medical devices are being used in the healthcare environment, or by a patient in a home setting. It can also be useful to learn how patients are interacting with clinicians in certain situations.

One of the emerging and interesting variations in observation is the use of wearable technologies to observe and understand true patient behavior. Wearable technologies are devices that are worn by individuals, and that monitor the person's movement, heart rate, sleeping pattern, and more. There are many different devices that can be worn on wrists, as vests, on footwear, and more. Contract research organizations (CROs) are already using them to monitor participants in research studies, and the US Food and Drug Administration (FDA) has issued guidance presenting regulatory views on the use of mHealth technologies in clinical trials. The increasing prevalence and familiarity of these devices among consumers, as well as the increase in accuracy, has attracted interest from many of the big players in this space.

Wearable technologies automate data collection as the researcher does not need to question participants. He can simply monitor their behavior through the device. They also produce much more precise data as participants do not have to estimate answers and there are no issues relating to recalling answers; the researcher has access to the exact data precisely calculated by the device, which he can obtain through an internet connection. As such, this method mitigates one of the key disadvantages of observation, which is the time necessary to observe participants.

The increasing use of consumers' wearable fitness trackers can also be a boon for companies wishing to understand consumer patterns. And there is no telling what the future holds in this space. Imagine using Google Glasses—a head-mounted display shaped in the form of glasses that can record activity—to monitor a person's shopping experience and later analyzing shopping patterns. Or feasibly, a company could use 3D virtual tools to enhance focus group participants' experiences by simulating environments. The emergence of and rapid changes in technology will play a role in how market research observation occurs, and how data can be collected.[20]

2.5.7 Mystery Shopping

Mystery shopping is a market research activity where trained individuals are mandated to evaluate the quality of service or the compliance with regulation. To do so, these individuals are sent to experience the organization from the outside, as an everyday customer. When performed, the locations that are the subject of the mystery shopping activity are not aware that they are being evaluated. This allows the head organization to better appreciate how it is performing, and to adjust processes and internal procedures accordingly.

The mystery shoppers act as much as possible like everyday consumers, asking questions, purchasing products, or behaving following certain scenarios, and then reporting their experiences back to the market researcher coordinating the effort. Mystery shoppers can investigate a number of items, such as:

- The customer experience: Number of employees, ease of service, and attitude of employees.
- The facility: Cleanliness, ease of navigation.
- The purchasing experience: Time spent to purchase product, and ease of transaction.
- The post-purchasing experience: Ease of returning product or using the return policy.

While it is possible for you to conduct mystery shopping in-house, it is more efficient to hire third-party companies or freelancers to engage in mystery shopping. They are more likely to reflect the point of view of an unattached observer, unlikely to nuance their observations with their own insider knowledge, and less likely to be recognized by staffers being evaluated.

Finally, it is possible to use mystery shopping to evaluate competitors, but a certain number of ethical guidelines should be respected. First, the mystery shopper should not record competitors' employees with a recording device as these employees have not given their consent. In addition, the length of time spent should be the equivalent of a normal market transaction (so as to not waste competitors' resources unethically). Finally, the evaluation scenario should not require a follow-up call from the company being observed, unless this is a normal part of this transaction. Globally, if the mystery shopper is asking questions that a shopper usually asks (e.g., price, quality, or availability of goods) and if no confidential business information is sought or revealed, then the common-sense rule of "no harm, no foul" applies.[21] Even then, if these guidelines are respected, some experts believe that mystery shopping competitors is walking into a grey area of market research, putting you at risk of liability.

2.5.7.1 Use in Life Sciences

Mystery shopping is quite useful for mature organizations that are already generating sales (or offering a service) and need to gain a deeper understanding of their customer experiences. It is also quite popular in healthcare facilities, as well as in new healthcare mobile application technologies.

In mature organizations (generating sales), mystery shopping is useful to evaluate the customer process and identify pain points in the inquiry, purchasing, and maintaining phase. If you have issues generating recurring purchases, consider a mystery shopping initiative to evaluate your own internal processes.

Healthcare facilities utilize mystery shopping to evaluate the level of service they are offering. It provides insight into the patient's experience and identifies gaps in the service chain. As such, a typical healthcare mystery shop could include:

1. Scheduling an appointment with a healthcare provider
2. Visiting the healthcare center

3. Completing a medical consultation
4. Reporting back on the patient experience

Also, mystery shops are often used by technology companies developing digital applications. For example, some firms will engage mystery shoppers to download, install, and register their app and then report on ease of use, issues during installation, and overall feedback.

Finally, some telemedicine apps will use mystery shoppers in an effort to evaluate the application's ease of use and solidity. They will use mystery shoppers to engage with clinicians, enquire about medical conditions, and even have the mystery shoppers use the app to purchase medications [either over-the-counter (OTC) or prescription] online. Typically, various elements are being verified, from the customer experience to quality of service all the way to adherence to regulation (is the customer able to purchase restricted products, or is regulatory compliance enforced?).

2.6 Artificial Intelligence and Market Research

I need to start this section with a simple reality check: AI is evolving so quickly that looking back at what we knew just 1 year ago feels like studying a different era. Assumptions we took for granted in 2024 were already outdated in 2025. That is the inherent limitation of the written word: By the time you read this, some of what follows may already feel inaccurate. So, I am keeping the discussion high-level and focused on durable principles rather than technical details.

In late 2025, Open AI published a paper[22] openly acknowledging something many researchers had already observed. No matter how perfect the data, no matter how we frame our requestion, current AI models are always capable of generating hallucination. And as of late 2025, there is still no reliable way to predict when it has generated something subtly wrong yet in a completely confident voice, beyond double-checking everything. That is a problem when your objective is to uncover the market's reality.

Here's the core misunderstanding about AI: Many people assume it reads, thinks, and "understands." We speak to it like a collaborator, so we start seeing it as intelligent, and even sentient. The branding certainly does not help. But despite the name, there is no intelligence in artificial intelligence. These systems are probability engines, predicting the most likely sequence of words. They do not "know," they generate information. And when they do not know, they guess, which at this time, is impossible to distinguish without the proper expertise.

So, what the study shared is that this guessing behavior is not a flaw, it is a part of the way these engines are designed. Since training models reinforce "give a complete an answer" behavior not "acknowledge uncertainty," they keep guessing. Which means that currently, market research teams cannot simply outsource fact-finding to an AI agent and assume the citations will hold up. I tested this myself while researching companies working on gold nanoparticle oncology. AI gave me a solid looking list of candidates, but a quick internet check revealed a company that had shut down (but looked alive online), and another that had been acquired years ago with outdated web assets still floating around. On the surface, everything looked right but after some market research, everything was wrong. Without a human in the loop, I could have grabbed this list and shared it with a client without a second thought.

So, if we cannot rely on AI to gather facts, what do we do with it when it comes to data gathering? Well, AI is being explored as alternative to conduct one-to-one interviews. The appeal is certainly there. Not only in terms of cost, but also in terms of potential consistency, from one interview to the next. Also, the possibility of having AI analyze data sets real time and using that data to conduct even more complex interviews is certainly enticing. They offer advantages in terms of scalability, cost effectiveness improved efficiency.

However, broader societal acceptance and resistance exist. Not everyone is overly excited about being interviewed by an AI agent. In an eastern European study, researcher found three types of participants: Open skeptics, technology skeptics and technology friends. While 29% of participants contacted where willing to participate, 22% were neutral and 49% were reluctant to be interviewed by an AI agent. This could potentially lead to sampling issues, where a certain population self excludes from a study, creating an incomplete data set when trying to complete your market research. Factors that reduced willingness to participate with an AI agent included lack of trust in the technology, not wanting robots to take human jobs, concerns overs data privacy and some would just rather talk to a human.[23]

But for now, as of 2025, if you do use AI to research information, treat it as a short list to investigate further, not as an exact data set.

2.7 Closing Words on Ethics and Primary Market Research

Throughout this chapter, we presented and discussed many different tools that a market researcher can use to collect data from participants. Sometimes, these participants will be aware of the marketing research effort, sometimes not. In any case, there are a few ethical considerations that a market researcher must remember.

First, it is very important to be honest when you collect data, identifying yourself and describing why you are collecting the data. Misrepresentation is a huge challenge in data collection, and it is tempting to do so to ease the data-gathering process. For example, some researchers will pretend to be a potentially interested party or pose as a student gathering information for a school project. This is clearly unethical. Instead, efforts should be spent identifying targets that have the information and are more likely to want to share it, such as academics, technology vendors, advertising agencies, and journalists.

Also, if the participant asks, you should be able to answer what the project is for. Not being able to give answers on the nature of the project gives the appearance of spying, not market researching. Companies engage in market research activities all the time, and you should be able to answer these types of questions truthfully. If you cannot, then ask yourself if this is a project that you should be doing.

It is very easy for a market researcher to influence the participant's response. Asking leading questions can cause a participant to answer in a specific way. Even agreeing with a participant rather than impartially acknowledging their answer can influence the participation, taint future responses, and fundamentally making it unusable. One has to remember that leading participants to answer a certain way might let you find the answer you want to hear, but this will not reflect the real market's appreciation of your product: Would you not rather find out the real market requirements during the market research phase, rather than shaping market research to fit your preconceived ideas, and then failing during commercialization? *While market research shapes our vision of the market, it does not change the true nature of the market.*

Furthermore, respect the confidentiality of the participants. If you have given them the assurance that you will protect their responses, be ready to do so. If you believe you cannot assure the participants' confidentiality, or if you do not intend to do so (by sharing the results to other stakeholders), be upfront so that participants can have the option of opting out. As an example, a few years ago, a client had contacted me to conduct a survey of life science companies. Participants were assured anonymity and answered truthfully about such as cash on hand, burn rate, and strategic priorities. After compiling the data, the client became interested in some of the information collected and wanted access to granular data so that he could share it with other stakeholders. I refused. While I lost a client, I maintained the privacy I had assured participants.

Finally, *primary market research is not a commercialization activity.* Engaging participants in market research and then trying to sell a product midway during data collection undermines market research. In some cases, a client might express interest in a product you are researching. When this happens, my approach is to ask the participants "You seem to have some interest in product X. Would you like me to refer you directly to Company X as an interested party? Do you accept that your coordinates be shared directly with the appropriate person?" As such, the participant is authorizing you to share his information and his interest. You are serving both parties, and with consent, are relieved of other obligations you might have (e.g., keeping the data anonymous). In addition, by engaging with the client this way, you are validating the true intent of the participants: A client willing to share his own information to be directly followed up is literality putting his money where his mouth is.

Notes

1 Miller, G. A. 1956. The magical number seven, plus or minus two: Some limits on our capacity for processing information. *Psychological Review*, 63(2), 81–97.
2 Henning, J. 2016. *The Qual Sandwich Doubles Down on the Survey.* Insight Associates. https://researchscape.com/blog/the-qual-sandwich-doubles-down-on-the-survey. (Last accessed January 20, 2026).
3 An interesting theory proposes to ask negative questions when a participant's privacy is a key concern. It argues that negative questions allow participants to keep the target data undisclosed by asking them, instead, to make a series of decisions with the information in mind. In this way, for example, the frequency of drug use can be calculated without respondents admitting to using any drugs. For more information, I suggest looking at "Surveys with Negative Questions for Sensitive Items" by Fernando Esponda and Victor M. Guerrero (2009).
4 Effective emoji use for public agencies" is: Kagarise, W. 2024. Effective emoji use for public agencies. *Journal of Digital & Social Media Marketing*, 12(2), 141–153.

5 Kagarise, Effective emoji use for public agencies.
6 Schmidt, M., Barth, A., & Blasius, J. 2025. Effects of mode and incentives on response rate, sample composition, and costs—experience from a self-administered mixed-mode survey of movers. *Survey Methods: Insights from the Field.* https://surveyinsights.org/?p=?p=19802 (Last accessed January 20, 2026).
7 Hawa, S., Bane, S., Kinsler, K., Rector, A., Chaichian, Y., Falasinnu, T., & Simard, J. F. 2024. Impact of incentives on physician participation in research surveys: Randomized experiment. *JMIR Formative Research*, 8, Article e54343. https://doi.org/10.2196/54343.
8 Cook, D. A., Wittich, C. M., Daniels, W. L., West, C. P., Harris, A. M., & Beebe, T. J. 2016. Incentive and reminder strategies to improve response rate for internet-based physician surveys: A randomized experiment. *Journal of Medical Internet Research.* Eysenbach, G., ed. https://doi.org/10.2196/jmir.6318.
9 Chen, J. S., Sprague, B. L., Klabunde, C. N., et al. 2016. Take the money and run? Redemption of a gift card incentive in a clinician survey. *BMC Medical Research Methodology.* https://doi.org/10.1186/s12874-016-0126-2.
10 The Market Research Society (MRS) is a UK-based market research association for practitioners. It produces a series of guides on a number of topics interesting for market researchers. Please visit their website (www.mrs.org.uk) for more information.
11 Kelly, D., & Rupert, E. 2009. Professional emotions and persuasion: Tapping non-rational drivers in health-care market research. *Journal of Medical Marketing: Device, Diagnostic and Pharmaceutical Marketing,* 39, 3–9.
12 Kerasidou, A., & Horn, R. 2016. Making space for empathy: Supporting doctors in the emotional labor of clinical care. *BMC Medical Ethics*, 17, 8.
13 Some moderators encourage the use of dyad focus groups (two individuals) instead of triads. These groups can focus even more on product testing and feedback, simulating a small work group in an organization. The downside is that dyads offer a limited source of information in terms of richness of context.
14 Elliott, M. N., Brown, J. A., Hambarsoomian, K., et al. 2024. Survey protocols, response rates, and representation of underserved patients: A randomized clinical trial. *JAMA Health Forum,* 5(1), e234929. https://doi.org/10.1001/jamahealthforum.2023.4929 (Last accessed January 20, 2026).
15 Media Logic. 2016. *Medicare Marketing Insights: What Direct Mail Designs Do Seniors Prefer?* March 23, 2017, https://www.medialogic.com/blog/uncategorized/medicare-marketing-insights-direct-mail-designs-seniors-prefer/ (Last accessed January 20, 2026).
16 Cunningham, C. T., Quan, H., Hemmelgarn, B., et al. 2015. Exploring physician specialist response rates to web-based surveys. *BMC Medical Research Methodology.* https://doi.org/10.1186/s12874-015-0016-z (Accessed January 20, 2026).

17 Pit, S. W., Vo, T., & Pyakurel, S. 2014. The effectiveness of recruitment strategies on general practitioner's survey response rates—A systematic review. *BMC Medical Research Methodology*. https://doi.org/10.1186/1471-2288-14-76 (Accessed January 20, 2026).

18 Suppan, M., Suppan, L., Beckmann, T. S., Samer, C. F., & Savoldelli, G. L. 2024. Enhancing response rates in web-based surveys: The impact of direct participant contact. *Healthcare*, 12(14), 1439. https://doi.org/10.3390/healthcare12141439 (Accessed January 20, 2026).

19 Pew Research Center. 2016. *Smartphone Ownership and Internet Usage Continues to Climb in Emerging Economies*. http://www.pewglobal.org/2016/02/22/smartphone-ownership-and-internet-usage-continues-to-climb-in-emerging-economies (Accessed January 20, 2026).

20 Robbins, R. 2016. Customers flocked to this futuristic medical tricorder. Now they're crying foul. *Stat*. https://www.statnews.com/2016/12/14/scanadu-tricorder-crowdfund (Accessed January 20, 2026).

21 Craig, E. 2007. Mystery shopping. *Competitive Intelligence Magazine*, 10(3, May–June 2007).

22 Kalai, A. T., Nachum, O., Vempala, S. S., & Zhang, E. 2025. *Why Language Models Hallucinate*. arXiv preprint arXiv:2509.04664. https://arxiv.org/abs/2509.04664 (Last accessed January 20, 2026).

23 Danó, G., Kovács, S., & Surman, V. 2025. Challenges and opportunities of AI in market research: Virtual interviewers. *Journal of Region, Society and Economy*, 13(1), 32–46. https://doi.org/10.14513/tge-jres.00413. https://tge.sze.hu/tge/article/view/413 (Last accessed January 20, 2026).

Chapter 3

Secondary Research

I am saying nine instead of ten given I am making it up and ten sounds too perfect and reality is never that neat.

—Beau Miles

Secondary research is the collection and collation of information that is already published and publicly available. Also called desk research, it is frequently conducted to explore a topic before engaging in more expensive primary research, or to quickly gain a summary understanding of a topic without engaging too many resources.

The advantages of conducting secondary research (versus primary research) are numerous. First, secondary research is considerably less expensive than primary research. Also, it is considerably faster to complete as it does not depend on third parties (e.g., recruited participants, organized focus groups, or enrolled survey participants) to obtain the information, and the sample size of third-party research reports is often quite considerable. Furthermore, due to scope and reputation, third-party information often has more perceived authority and impartiality than "in-house" research. A statement such as "MarketsandMarkets estimates that the global cell-based assays market, valued at US$17.36B in 2024, is forecasted to grow at a 9.1% CAGR, reaching US$18.134B by 2025 and US$25.77B by 2030."[1] has more credibility than most in-house estimates. Finally, as mentioned at the beginning of this section, secondary research is very useful to orient and define primary research: a researcher will often start a market research project by doing a quick market review to identify

DOI: 10.4324/ 9781003731726-3

some of the main trends and concerns before diving into deeper primary research.

There are some disadvantages to secondary research. First, the information is not always personalized to an organization's requirements, and it is quite difficult to find data for emerging fields: reports on nanomedicine are plentiful, but developing a specific application merging nanomedicine and artificial intelligence means that recognized secondary research will be quite scarce. This means that the organization will need either to (a) extrapolate from generalist research, or (b) conduct primary research. Also, the data might be outdated, limiting its usefulness. Finally, most of the time, the original data used in secondary data is unverifiable: it is quite difficult to spot errors in data collection or dispute the way that data was analyzed in a consolidated report.

There are two types of secondary data: External and internal. External secondary data is information gathered from outside the organization. This includes anything from government statistics to media sources. Internal secondary data is data that the organization is generating itself. It could be data collected from customer feedback, accounting and sales records, or employee experiences. The bulk of our suggestions are related to internet sources for ease of use, convenience as well as cost.

Finally, we make a distinction between active and passive secondary research: Active secondary research takes place when the researcher is actively searching for information, while passive secondary research is the use of tools and software to automate data collection.

3.1 Active Secondary Research

Active secondary research takes place when the market researcher is dynamically searching for information. In this section, we will be going over some of the most popular and pertinent sources of data available. All the suggested sources are web-based. The internet has grown to be by far the most important resource for searching for and identifying secondary research information: Even most magazines and newspapers have made their archives at least partially available online.

While there are some ethical considerations for what information the researcher can use, a simple rule of thumb is that any information made available to the public is fair game for collection and review. If the method used to obtain that information is not commonly available to the public (e.g., using a former employee's password to access a restricted website area), then it is not only unethical but also most definitely illegal.

In the following section, we will be going over a variety of internet data sources that a market researcher might wish to investigate when conducting secondary research, and will conclude with a few tips to increase the effectiveness of Google-based internet researches.

3.1.1 Government Data

Government agencies generate large bodies of information that can be used by researchers. Most of this information is free to use and can be useful at the start of a research project. Government data is usually statistical in nature and is very useful when building marketing models or trying to understand the fundamental nature of a market. Table 3.1 provides a selection of government databases that could be useful to review during a market research effort in life sciences.

3.1.2 Public Company Data

Companies publish a significant amount of useful information online. Often, companies will do market research and will publish results as consolidated information in corporate documentation. As such, reviewing this publicly available data is another useful way to start your research effort.

One of the first places to search online is the website of a specific company with a technology or product akin to the one that is the focus of your research. Looking at data from other companies can help to estimate market size, growth, and segmentation. For example, while perusing a corporate presentation, you might find a sentence on market segmentation where the company states that it believes to be the second largest player in the market and estimates that it generates 30% of the total sales. This statement might be based on other secondary research or on internal data, so it is important to note the origin of the statement. Some of the documents that can be reviewed online include:

- Annual reports, which might also be available on the websites of central depositories, such as the Securities and Exchange Commission (for American companies, www.sec.gov) or SEDAR (for Canadian companies, www.sedar.com)[2]
- Company pitch decks or presentations at investor's forums.

Table 3.1 Government Databases for Secondary Research

Organization	*Website*	*Why It Matters*
WHO Global Health Observatory	www.who.int/data/gho	Core global health indicators (disease burden, mortality, and risk factors)
World Bank Open Data	data.worldbank.org/	Socioeconomic and development indicators (GDP, health spending)
WIPO Patent Scope	patentscope.wipo.int/	International patent search (complements USPTO and EU patent tools)
OECD Health Statistics	stats.oecd.org/	Cross-country health system indicators (spending, beds, and physicians)
Data.gov	www.data.gov	Central catalog of US government datasets
US Census Bureau	data.census.gov	Demographics, income, and geography
National Center for Health Statistics	www.cdc.gov/nchs/	Core US health statistics source (morbidity, mortality, and utilization)
CDC FastStats	www.cdc.gov/nchs/fastats/	Quick-access health stats by topic
ClinicalTrials.gov	https://clinicaltrials.gov	Global clinical trial registry
US Patent and Trademark Office	www.uspto.gov	US patent search
Statistics Canada	https://www.statcan.gc.ca	Demographics, economics, health, and social data
Health Canada	health-products.canada.ca	Approved drugs (and certain device/product registries
Canadian Institute for Health Info	www.cihi.ca	Health system utilization, hospital, physician, and cost data
Eurostat	ec.europa.eu/eurostat	EU-level economic, demographic, and health indicators

(Continued)

Table 3.1 *(Continued)* **Government Databases for Secondary Research**

Organization	*Website*	*Why It Matters*
EMA—European Medicines Agency)	www.ema.europa.eu/	Centralized EU drug approvals
EU Clinical Trials	euclinicaltrials.eu	Clinical trials in the EU
NICE (UK)	www.nice.org.uk	Guidelines for England
NHS Digital/NHS England Data	digital.nhs.uk	Utilization, activity, waiting lists, prescribing data

- Press releases, marketing collateral, product brochures, and white papers.
- Video product presentations (take the time to visit YouTube to verify if the company has a YouTube channel, and if it has published online content on it).

Once these documents are identified, be sure to download them and save them to your own local drive. Companies can change or update their content online without warning, and the information you identified might not be available any longer when you return to the website. Taking a screenshot of important information can be another way to ensure that the data is not lost following a website update.

3.1.3 Print Media

Print media sources are published on a regular schedule by specialized companies. These include magazines and newspapers and include both their printed/physical format and their internet counterparts.

Generalist newspapers and publications, while interesting, are of limited use when conducting specialized market research. Although it is possible to get general information from them (e.g., trends, government policies and orientations, or industry summaries), print media market research in life sciences will focus mostly on specialized content publishers, mostly in the form of trade magazine publishers.

There are a number of trade magazines that are published on a regular basis that are of use to market researchers. Most of them are free, and access to their archives is public, although some publishers monetize their archives.

Table 3.2 Publications of Interest in Trade Media for Life Sciences

Publication/Media	*Website*	*Focus/Why It Matters*
Genetic Engineering & Biotechnology News	https://www.genengnews.com	Bioindustry news and detailed innovation trends
FierceBiotech	https://www.fiercebiotech.com	Biotechnology deals, funding, and partnerships
FiercePharma	https://www.fiercepharma.com	Pharma approvals, policies, and commercialization moves
BioProcess International	https://www.bioprocessintl.com	Biomanufacturing, scaling, and Controls issues
Contract Pharma	https://www.contractpharma.com	Pharma outsourcing, CDMOs, and supply chain dynamics
MD+DI	https://www.mddionline.com	Medtech and diagnostics news
BioCentury	https://www.biocentury.com	Strategic intelligence, valuation, and deal flow
MedTech Dive	https://www.medtechdive.com	Medtech business and regulatory tracking for your competitive analysis

Table 3.2 lists some of the main publications that market researchers can use during their market research efforts. You should also investigate whether publication publishes specialized inserts or special issues on focused topics, as these also provide highly valuable market data.

3.1.4 Trade and Industry Groups

Trade and industry groups are organizations representing multiple entities—including private companies, government agencies, universities, and consultants—within a common commercial activity sector. Some of the larger associations produce (or sponsor) reports relevant to their industry, which can be of use during market research. One of the advantages of the reports produced by these organizations is that they are third-party reports and are relatively unbiased; however, the reports are not always freely available and are sometimes reserved for members.

Associations are also useful for identifying participants or targets for further research as their members are routinely listed on their websites. This can give the researcher a good starting point when researching a specific technology area, or if the researcher's research focus is in a specific geographical region. Finally, many trade associations collect information from their members in the form of surveys and then share the results at large in a consolidated format, producing reports on key trends and issues of concern. The most important life science trade organizations are listed in Table 3.3.

Table 3.3 Main Life Science Industry Groups

Association	*Website*	*Focus/Why It Matters*
Biotechnology Innovation Organization (BIO)	https://www.bio.org	Leading biotech advocacy group in the U.S.
PhRMA	https://www.phrma.org	U.S. research-based pharmaceutical industry
Association for Accessible Medicines (AAM)	https://accessiblemeds.org	Generic + biosimilar manufacturers
AdvaMed (Advanced Medical Technology Association)	https://www.advamed.org	Largest U.S. medtech association
Medical Device Manufacturers Association (MDMA)	https://www.medicaldevices.org	Represents smaller and mid-size medtech innovators
MedTech Europe	https://www.medtecheurope.org	Main device and diagnostics industry voice in Europe
EuropaBio	https://www.europabio.org	European biotech industry association
BIOTECanada	https://www.biotech.ca	Canada's national biotech/pharma association
American Hospital Association (AHA)	https://www.aha.org	Represents U.S. hospitals and health systems

3.1.5 Scientific Publications

Scientific publications are a highly useful source of information when researching market trends and future technologies. They can also be very useful to identify key opinion leaders for future research efforts—such as interviews and focus groups—and they supply valuable information to benchmark technologies. While there are many publications available, the best approach is to target hubs that consolidate articles (Table 3.4). Note that market data in scientific journals can be dated as the publication cycle can be longer.

3.1.6 Market Research Firms

Large market research firms publish syndicated reports on a regular basis. They collect data from major participants in an industry, consolidating the data and publishing it so that it can be purchased by all interested parties. The reports can be expensive but can be worth it for the time saved by purchasing a completed product (rather than engaging in a costly primary research operation). Some companies also offer report customization for its bigger clients, enabling some level of personalization. For some reports, it

Table 3.4 Main Scientific Article Hubs for Market Research

Source	*Website*	*Focus/Why It Matters*
PubMed Central	https://pubmed.ncbi.nlm.nih.gov	The primary biomedical literature index
Google Scholar	https://scholar.google.com	Broad scholarly index including journals, theses, and patents
Europe PMC	https://europepmc.org	Similar to PubMed but with added access to **patents, NHS guidelines**, and more EU-centric research
Springer Nature	https://link.springer.com/	Major open-access publisher of clinical and biomedical research
Directory of Open Access Journals	https://www.doaj.org	Curated list of validated open-access journals

is also possible to purchase a single chapter, making the purchase much more affordable. Finally, executive reports on these syndicated reports are sometimes available at large, so it is possible to collect some data that can be used for data triangulation.

3.1.7 Blogs

Blogs are websites or publication streams maintained by individuals and companies to publish information in a more interactive format. The objective is to generate discussion between an organization and its consumers, or between an expert and people interested in a topic. It is also heavily used by consumers and advocacy groups to express opinions, both positive and negative, related to a topic of interest.

Blogs used to be one of the top trusted sources of information due to their personal nature. A Technorati survey found blogs to be one of the top five trusted sources of information on the web—following news sites, Facebook, retail sites, and YouTube.[3] However, the fact that they were so trusted led to a deterioration in the quality of information being shared on them. It was quite easy for bloggers to manipulate (or be manipulated), so the information they share should be taken as an information lead, rather than being positioned as a definitive fact. As Ryan Holiday mentions in his book *Confessions of a Media Manipulator*, "The economics of the internet has created a twisted set of incentives that make traffic (to the blog) more important—and more profitable—than the truth."

Since the first edition of this book, blogs have continually declined, and they have slowly been replaced by podcast as means of engaging client bases. While people debate the usefulness of blogs when so many ways to push information to consumers exist, blogs still have some very important characteristics compared with alternatives (social media websites e.g., LinkedIn or using a Facebook Feed). The most important of these differences is that as blogs that blogs located directly on an organization's website allow for more detailed posts and have more longevity, as they are available for as long as the organization wishes them to be online.

Researching blogs is a useful way to identify trends, consumer reactions, and key opinion leaders. Blogs from advocacy groups are especially useful to identify concerns from patients, and blogs from KOLs are useful to identify upcoming trends, but care should be taken to measure these opinions against other sources of information.[4] In the end, people have to reframe the roles of blogs. As Artemisa Forbes noted on one of my LinkedIn posts, blogs are not being replaced by podcast since they serve different research functions. Technical blogs remain critical for niche B2B expertise, especially in industries where practitioners share knowledge (geotechnical, environmental compliance, and industrial processes). While podcasts are useful for discovery and relationship-building, blogs remain critical for references and credibility.[5]

3.1.8 Social Networks

For this book, "social networks" are defined as dedicated websites or applications where users with a common objective gather and participate in discussions. These discussions create a network of social interactions, as users share messages, comments, information, opinions, experiences, and more.

Currently, the most useful social network in a business context is LinkedIn. It is useful to identify interview targets, thought KOLs, and groups around specialized topics. LinkedIn content—which can now include newsletters and articles—has replaced traditional blogs for many B2B though leaders, as they generate more traffic and are easier to maintain. It is also a useful to share your profile and find new projects.

Other popular social networks are as Facebook and Instagram (which focus more on social interactions), as well as X and Bluesky (which are used for commenting and quickly diffusing information and opinions). However, there are hundreds of social networks catering to specialized needs for niche communities. A highly interesting one for researching trends in life sciences is the PatientsLikeMe (www.patientslikeme.com) social network, a website that bills itself as a "free website where people can share their health data to track their progress, help others, and change medicine for good."

Social networks are a useful way to measure the customer's pulse on a topic. Analyzing the comments found on social network pages allows the researcher to understand how people view a brand, or their perception of a specific topic.

The popularity of social networks has led many companies to set up and maintain pages on social networks as well. Some pages are for the company in general, while others are specific product pages. When reviewing these pages, it is possible to collect and collate customer feedback on these specific products.

For example, say you were developing a carrying device for glucose monitoring devices. Using Facebook, you would find the Medtronic

Diabetes Facebook page.[6] Reading through the comments, you would find information such as:

- *I would like to know why the pump clips are made so cheaply now 😞😞! I am considering gluing it to my pump at this point because I am so tired of it snapping off with the slightest bit of a bump and no it is not a default in my clip itself I had two sent to me and both are just as bad as one another . . . the quality of this product is going down but the prices seem to keep going up and it is truly sad . . .*
- *Both of my clips have broken in the last few months. One of them broke while I was sleeping!*
- *I stopped using those plastic ones for that reason.*

Through further research, you might be able to identify an opportunity for a more sturdier carrying solution or something completely different from the current technology.

A good model for analyzing data gathered on social media is using a slight variation of the Five Ws (Who, What, Where, When, **and** Why).

- **What happened?** What was the situation that caused the posters to post messages on social media?
- **Who is involved?** Who is posting the message? Who else is involved? Is the poster posting for somebody else?
- **Where was the comment posted?** Where did the poster post his comment? Did he append his comment to a specific post, or did he start a new conversation?
- **Why did it happen?** Why did he post a message? Why did he use social media instead of a more conventional method (e.g., calling customer support)?
- **How can this problem be solved?** What did he hope to accomplish? What are the solutions that would resolve this situation?

Reddit is another good source for finding informal information. Is acts as both a social media platform and online forums where users can share content and opinions. It has been harvested to be used as the building blocks of many AI engines as it is perceived as reflecting the common's man perception. You can use it to quickly research topics related to a technology or a trend. For example, on Reddit in November 2016, someone started a discussion called "What are your thoughts on the Bayer x Monsanto merger?"[7] Some of the feedback included:

- *Could be something to do with weed legalization. I know Monsanto has wanted to grow large scale pot farms, and that means a lot of drugs to process. Bayer has a history of processing drugs especially for the pain relief sector. I have no proof of this whatsoever. But no one else commented and as a weed smoker this merger turned some lights on for me.*
- *That conspiracy theory just never dies. They have stated (over and over and over) that they have no interest in marijuana.*
- *Fun thought, but extremely improbable bumping up against impossible. By this logic, why not get into the tobacco market? Growing commodity crops, which is what marijuana would be if legal, is not in the interest of a chemical company. Bayer wants to get into agricultural chemicals. More so than they already have and the easiest way to achieve that when you are sitting on mountains of cash are to buy someone.*

You must be clear, these are rumors, not facts, akin to something a researcher might overhear in a café or a convention. And without data triangulation, these are random statements, and should not be taken at face value. Nonetheless, what they do provide is a potential clue, and a research lead.

Reddit is a place where people familiar with the industry converge to engage in discussions. It is possible for a researcher to use it to test a hypothesis, or to discreetly ask participants a question to get more potential information.

3.1.9 Search Engines: Tips and Tricks

Searching through the internet using a web research engine is usually the first step of your secondary research. Here are a few tips and tricks to make the research effort more efficient:

- *Look to the past:* Sometimes, a market researcher will be looking for something specific but will conclude that the information is no longer available online. It might be an old press release that a competitor has pulled from their website, information on a previous partnership that was quietly ended, or specifications on discontinued products, for example. In these cases, the website www.archive.org (an internet archive non-profit digital library offering free universal access to all) is especially useful. Archived in their public database are historical web snapshots of a company's website, which can include pages, attachments, and more. While a researcher might not have access to each version of a company's website, there are often several snapshots taken throughout the year, enabling the researcher to identify key information that has been removed online.

- *Look for corporate web DNA:* I originally found this technique referenced by Leonard Fuld in *The Secret Language of Competitive Intelligence.* It is based on the concept that every organization develops their own brand of corporate speak, or pattern. It is akin to corporate web DNA. Fuld defines it as "a unique pattern of words and phrases that form the substance of a company's website, its press releases, and its advertisements."[8] As such, if the researcher can identify a group of unique words or jargon as potential corporate DNA, he can then proceed with searching the web using the aforementioned terminology, grouped between two sets of quotation marks. As an example, using Medtronic's "Transforming technology to change lives" slogan to search the web brings up a series of white papers (old and new), job offers (both current and expired), as well as customer testimonials.
- *Go beyond Google:* If you are not finding the information you need, you can try using another search engine to obtain different search results. Some of the interesting alternative search engines include:
  - Bing (www.bing.com) (which is reported to have a better video search option), DuckDuckGo (which values privacy) and Kagi (a private search option).
  - Board reader (www.boardreader.com) (which specializes in user point of views by searching through forums, message boards, and Reddit).
  - Slide Share (www.slideshare.net) (a cornucopia of PowerPoint presentations, slide decks, and webinars from past conferences).

3.1.10 Power Up Google

Google is the starting point for many research projects. As of 2025, Google accounts for over 90% of the global desktop search engine market share, as its simplicity of use and search algorithm for ranking websites in order of importance enables researchers to quickly find the information they are looking for.

When searching on Google, remember that the search engine ranks the first word as slightly more important than the second, and the second slightly more important than the third, and so on. Hence, searching for "North America vaccine trends" and "vaccine trends North America" will generate slightly different results.[9] There are several powerful options available when searching with Google. Some of the most useful ones are shown in Table 3.5.

Table 3.5 Tips for Searching with Google to Increase Results

Tips	*Examples*
Use the *similar* option under a search result to have Google Search look for similar web sources.	fig_003.tif
The " ~" punctuation mark makes Google search for synonyms to a specific word	Searching for ~neuroscience will search for neuroscience as well as neurophysiology, neurobiology, brain, and neurology
Include words in your search by using the *addition sign* before words or phrases, or exclude words from your search phrase by using the *minus sign*	Searching for "biosimilars+Asia" will bring back every search result with biosimilars and Asia. Searching for "biosimilars – Europe" will bring back every search result with biosimilars that excludes Europe
Restrict your search to a specific domain by using the *site* command; to get results from multiple sites or domains, combine with the *OR* operator. Reddit is a very popular site command as well.	To find information on Aspirin published specifically in *The New England Journal of Medicine* or from the FDA, you would type: *Aspirin site:eejm.org OR site:fda.gov*
The *filetype* operator lets Google search specifically for exact types of files, which is quite useful when looking for market data	To find some PowerPoint slides with information on market share for cardiology devices, you would type: *Cardiology devices "market share" filetype:ppt*

3.1.11 The Future of Google

One cannot discus Google as a search engine and not discuss its increasing reliance on artificial intelligence. Anyone who has used Google—or any other engine for that matter—will be aware of this phenomenon: engine now generate a small paragraph or two, in response to the query, attempting to best answer while showcasing supporting sources. This is both a time-saver and a source of concern.

While it is quite useful to see it as a quick overview of a topic, all previous warning and discussion about data consolidation and hallucinations remain. As such, while the information shared might make structural sense, it might be biased or incomplete. But it is quite tempting to look at the summary, use its conclusions and not do our own research through the various sources that where identified. This is especially perplexing, as this underlies the overall existence of Google: if publishers do not get visitors directly to their page, there is no need to generate new content, new pages, new blog posts. Eventually, there is no need to upgrade your page, and the information gets stale. Over time, as Google culls information for free through news websites, company blogs and private users, what is the interest in generating that content?

This underlying structuring challenge will most likely shape how we look for information in the next 2–3 years, and what will be available.

3.2 Passive Secondary Research

Automated internet research tools are a boon to market researchers, as they automatically monitor and report on specific information topics. We will be going through some of the most interesting tools that researchers can use to automate their market research.

3.2.1 Google Alerts

Google Alerts is a service offered by Google. It is useful to automatically monitor a topic by setting up a search alert. Once set up, the Google Alert sends search engine search results by email to researchers as they occur, or on a predetermined basis in the form of a digest (once a day or once a week), at a predetermined time.

To create a Google Alert, the user only needs to go to the website www.google.com/alerts, type in their topic of interest, and customize the information feed requested (frequency, where the information will be collected from, the number of results wanted each period, and which email will be receiving the information). It is possible to edit an alert if needed, or delete the alert if it is no longer needed.

Google Alerts monitors major sites and news media for the topics selected and easily shares the information it collects across social media platforms such as Facebook and LinkedIn.

Some best practices in setting up a google Alert include:

1. Use precise words and "quotation marks" around search terms to increase precision and reduce noise from false search terms.
2. Setup search term parameters around frequency, source, language, and region to reduce false positives
3. Create multiple small search alerts rather than one large search alert to make it easier to prune words that are no longer necessary and make it easier to manage the search results
4. Implement a triage system by setting up a new dedicated search Gmail (to avoid having your main Gmail account flooded by google alerts).

3.2.2 Web Monitoring Tools

It is possible to set up specialized software and online search tools that monitor changes in specific web content (e.g., a competitor's website). These tools will alert you to any changes on a company's website as they occur, making it unnecessary to continually monitor key websites. Some software can also be configured to simultaneously monitor RSS feeds and newsgroups, making powerful all-in-one solutions. Some of the most popular software in this space includes Uptime.com, Datadog, and Pingdom.

3.2.3 Social Media Tracking

Several tools exist that are specifically designed to monitor social media. These tools search through popular platforms, such as Instagram and Facebook, and analyze generated content such as blogs and comments—to generate reports that can be used to identify underlying consumer trends. Some of the best tools in this space include: BrandMentions (Best for real-time brand monitoring), YouScan (Best for comprehensive visual analysis) and Buffer (Best for user engagement analysis). Services like Sprout Social and Meltwater are also useful to analyze sentiment regarding your brand.

3.2.4 Google Trends

Services such as Google Trends track comprehensive search results over long periods of time and are useful to identify underlying trends. As such, a researcher can uncover what the potential target audience is searching for online and analyze the information to gain insights on potential customers—especially if your organization is targeting end users).

Google trends will be useful to gauge overall interest in a topic by region, show the top queries related to the topic at hand, and try to identify related topics. It is a valuable tool when identifying other relevant topics when researching a subject. Google Trends can be very useful to research a competitor and get a quick image of its overall popularity. Some other uses can include:

- Researching the product features that potential clients find important by identifying what search terms are the most popular in relation to the search topic.
- Gauging consumer demand by searching which products or features generate the most consumer demand.
- Evaluating what consumers are searching for in competing products, by entering a competitor's name in the topics box and looking at search queries related to the brand.

3.3 Internal Secondary Data

Internal secondary data exists and is stored inside the organization. It is available exclusively to the organization and is usually generated and collected during normal business activities. Internal sources of data should always be investigated first because they are usually the quickest, most inexpensive, and most convenient source of information available.

Sources of internal secondary data include:

- *Sales data:* If the organization is commercializing its product, it has access to an invaluable resource, its own sales data. This data is already collected by the organization and organized in a way that is useful and extractable by a market researcher. Some of the data that the researcher can look at includes sales invoices, sales inquiries, quotations, returns, and salesforce business development sheets. From this information, territory trends, customer type, pricing, elasticity, packaging, and bundling impact can be inferred. This data can be useful to identify the most profitable customer groups and which ones to target in the future.

- *Financial data:* All functioning organizations have accounting and financial data. This can include the cost of producing, storing, and distributing its products. It can also include data on Research and Development (R&D) costs and burn rate.
- *Internal expertise:* Mid-sized organizations will often have inside expertise, that is, personnel who have been with the organization for a while. These people can be tapped and interviewed to get more information on past initiatives, products, lost customers, or any other topics. Frequently referred to as the organizational memory, they often have a wealth of undocumented knowledge that can be harvested. They might be aware of internally produced reports that might be of use, past projects, or failed product initiatives.

One of the weak points of internal data is inaccuracy, due to the fact it may be dated or the way it was collected. Also, while most of the time data can be ported from internal systems to market research data analysis tools, some legacy systems may make data conversion especially challenging. Finally, there are confidentiality issues: Some companies employing third-party researchers may hesitate to "open the books" and share either consolidated data or limited data sets.

3.4 Evaluating Your Secondary Research

Most secondary research, by its nature, is generated by a third party (the exception being the internal secondary data we discussed in Section 3.3). As such, it is important to verify its accuracy before using it. For this, we propose the use of data triangulation.

Data triangulation is the validation of data through cross-verification of two or more independent sources to ensure its accuracy. For example, to estimate market growth, a company could use multiple market reports, consolidating them with key opinion leaders' market growth estimates and information gathered from government websites. Data triangulation is very useful to strengthen the company's assumptions going forward.

Data triangulation is especially important when dealing with sources of information that are akin to rumors and hearsay: Information found on blogs, social networks, and discussion groups should never be taken at face value but rather used as leads for further research. It has become so easy to manipulate these types of media (either purposefully, or though lack of rigorous fact-checking), mainly because of the way social media generates revenue—more clicks on the web pages means more views, which means more revenue from advertisements. Hence, popularity is a much more important metric than the truth. The information found on social media should never be implicitly trusted.

Of note, it can also be interesting to triangulate primary and secondary data as a way to verify data accuracy. One way to do this would be to collect some key pieces of data and discuss them with key opinion leaders during an interview to get their opinion and perspective.

There is also a notion of expired data sources. With the information updating quite rapidly, you must be wary of the information you find. Statistical information is often expired after 12 months, while market trends can change in a few months or even weeks. For technology data, with rapid innovation, Information can be obsolete after a month. Competitive information in digital space can be experienced in less than six months, while heavily regulated industry data sets can be valid for one to two years. As such, contextualizing the "freshness" of your data is a key consideration.

Finally, when triangulating data, it is crucially important to make sure that the sources used to triangulate do not refer to the same initial source. Using a series of news articles and blog posts that all refer back to the same initial market report would not enable triangulation. Likewise, more media stories are found to be inaccurate, having implicitly trusted blogs and built stories on inaccurate information.

To illustrate the importance of triangulation, a few years ago, I was updating a market size (Total Addressable Market (TAM) and Serviceable Addressable Market (SAM)) report for a client. He had prepared his own model a number of years earlier, and was looking to update it as well as to perform an independent validation. But we had a challenge as my current estimates were much smaller than his previous estimates. As he believed his market was growing, he could not understand why my estimates were so much smaller. After some research, I found that one of the market estimates that he had used was based on a study that defined his market segment much more broadly than what he was targeting. As such, as my client had used a single point of reference as his starting point, his model grossly overestimated his market.

3.5 Using AI for Secondary Research and Current Limits

AI has been touted as a great tool for classical market research. Supporters claim that AI can accelerate market scanning and trend detection, particularly by aggregating data from multiple sources. It seems faster and more efficient and can seemingly find answer to the most complex of problems without interacting with a single person. However, there are two major issues with current AI research technologies: Data hallucinations and inability to properly evaluate information sources.

First, when an AI has insufficient information, it will hallucinate and generate the most probable answer, but not necessarily the right one. Why? Because training and evaluation procedures reward guessing over acknowledging uncertainty. As such, AI will manufacture reports that do not exist, companies that never existed or non-existing scientific literature to answer the question. For example, in a twist of irony, a Newfoundland government report commissioned on the use of AI in education, included at least 15 citations for non-existent articles or sources.[10] This is one of numerous examples of companies of using AI to generate a report that complied with a pre-existing conclusion, and where AI generated reference to support this hypothesis.

AI also tends to scratches the surface without digging deeper and not validating sources. As an example, I used ChatGPT and asked it "Are biolubricants a great alternative for train companies?" The answer was resoundingly yes, it is a great fit and is used extensively. Looking closer at the source it used, I found the AI had used four consulting companies' websites in the biolubricant space, and one market report. Not exactly the most objective analysis in the world, as it quoted vendors as legitimate sources.

As such, the risks of "garbage in, garbage out," are clear when AI outputs cannot be trusted. Inaccurate information can lead to flawed market assessments, incorrect pricing strategies or misallocation of resources. And if we are researching a topic, we know little about, it becomes very difficult to evaluate the truthfulness of a statement, and to evaluate how accurate a data is.

Nonetheless there are ways to use AI in market research. For example, AI are useful tools to explore a theme and brainstorm on a topic. They can suggest sites to visit on articles to read. Scite.ai can also be useful to validate information and citations. But, as of early 2026, we are currently limited on how we can use these tools for research at large.

3.6 Limits of Secondary Research

As mentioned in the first few pages of this chapter, secondary data has several issues that must be considered. There are issues related to availability, accuracy, trust, and bias.

First, there are issues of availability: not every topic will have secondary research data that exists and fits the needs of the researchers. The more specialized the topic, the more difficult it is to gather data. As such, careful consideration has to be given to how a topic has been defined. For example, if you are researching the generic manufacturing industry, and a report describes the market size, there could be issues with how the market was defined (does the author include biosimilars? Does the market study make a distinction between patented drugs and over-the-counter (OTC) drugs?). The researcher has to double-check the definitions used in the secondary data he chooses to use.

Second, there are issues of accuracy, or the lack of data verification. A survey that you generate by yourself is verifiable, as you are able to revise your numbers quite easily. And you can trust the data that you generate as you were the one who gathered it. But a secondary market report or a news article cannot be verified easily. You do not have access to the building blocks, and have to rely on the publisher's honesty and accuracy. If the author is biased, has been manipulated, or has made an error, the market researcher will be at the mercy of the data that he is using. Media references, especially those that are very close to social networks, should be used as information leads, not information sources.

Thirdly, secondary data may not be sufficiently detailed or have such a small sample size as to not be useful. Like checking a scientific article, checking the sample size on which the author is basing his conclusions is critical when evaluating data. A data reference point without sufficient background on methodology should never be used.

Finally, as the author of market research must be careful to keep his bias in check, it is always useful to check for any potential bias in the data source being used. Some organizations and authors have their own agenda, which would taint the quality of information being shared.

Notes

1 MarketandMarkets. 2025, April. *Cell Based Assays Market: Growth, Size, Share and Trends.* https://www.marketsandmarkets.com/Market-Reports/cell-based-assays-market-119917269.html (Accessed December 9, 2025).
2 Unfortunately, there aren't any consolidated sources for European and Asian companies, so you will have to search through national exchanges for the information you need. The first step will be to identify where the company is listed.
3 Technorati Media. 2013. *Digital Influence Report.* https://www.slideshare.net/slideshow/technorati-2013-digital-influence-report/16951048 (Accessed January 21, 2026).
4 "Trust Me, I' m Lying: Confessions of a Media Manipulator" (Portfolio Penguin, 2013), written by Ryan Holiday, is an interesting read on the economics and role of the Internet in how information is generated in social media.
5 Artemisa Forbes. 2025, October. *LinkedIn post.* https://www.linkedin.com/feed/update/urn:li:activity:7379164472482897920/
6 Source: www.facebook.com/medtronicdiabetes (Accessed December 25, 2016).
7 Reddit Forum. 2016. *What are Your Thoughts on the Bayer x Monsanto Merger?* https://www.reddit.com/r/biotech/comments/5i8x5a/what_are_your_thoughts_on_the_bayer_x_monsanto (Accessed December 27, 2016).
8 Fuld, L. 2006. *The Secret Language of Competitive Intelligence.* New York: Crown Business.
9 Search demonstration completed January 9, 2017.
10 Butler, P. 2025, September 12. N.L.'s 10-year education action plan cites sources that do not exist. *CBC News.* https://www.cbc.ca/news/canada/newfoundland-labrador/education-accord-nl-sources-dont-exist-1.7631364 (Last visited December 19, 2025).

Chapter 4

Analyzing Data

> *We found no evidence that the good to great companies had more or better information than competitors. Both had access to the same data sets, so the key is not the information but turning this information into something that cannot be ignored.*
>
> **—Caroline Florence**

In this chapter, we will provide an overview of data analysis. Our objective is not to turn the reader into a statistical analyst guru, but rather to give him a better sense of how to read the data he has collected, and some tools to better understand it. As such, we will be going over the basic elements of data cleanup, followed by some words on quantitative and qualitative data analysis, and closing with some barriers to effective data analysis.

An important point I wish to make is that market research is sometimes seen as a sequential activity, with each activity being completed before proceeding to the next one. But it does not have to be conducted this way, and it can be iterative. You should review your data during data collection. For example, you could look at your data when you have collected 50% of it to see if any trends are emerging, then again at the 75% mark and at the end to see any patterns are confirming themselves. You do not have to wait for the moment where you have all your data to start analysis, because looking at data during data collection is an opportunity to go look for secondary data or to question ongoing assumptions with new participants.

It is also possible to "jump" back and forth between the collection and analysis of data. For example, during initial data analysis, it is possible to

DOI: 10.4324/ 9781003731726-4

find out that some new research areas you had not uncovered or some of your data is insufficient, and a new data collection effort is necessary. The researcher should account for this possibility.

Case in point, a while back, a client had conducted a web survey to identify some trends related to salary and incentives in healthcare. He had collected an impressive sample (over 3,000 entries), and I was brought in to analyze the data. Nonetheless, during the initial analysis of the data sample, the demographic data revealed that the overall weight of U.S. East Coast respondents was too large compared with the other U.S. geographic regions. As such, my client had to proceed with a second data collection effort, this time focusing on the other regions, if he wanted to obtain a representative sample.

4.1 Initial Data Analysis

Once your data collection is completed, it is quite tempting to immediately start data analysis. After all, the objective of the project is to answer one or more market research questions, and once you have your data in hand, it feels as if the answers are a footstep away.

However, taking time to clean your data is crucial. This must be your first step before data analysis. We call this step the *initial data analysis* phase. It includes familiarization (where you gather and cleanup your data), codification (where you start labelling your data set) and classification (where you group your data sets into categories). While often undocumented in the final report, some researchers state that as much as 80% of the time allocated to the statistical analysis process is spent on data cleaning and preparation.[1] As such, it is essential that your research plan includes a budget (in both time and money) for this initial data analysis.

Note that at this stage, the researcher is not analyzing the data itself to find the answers to his research questions but rather focusing on the data set to evaluate its validity and adjusting it as needed (while documenting any adjustments that are made). This is a crucial step, especially if more complex data analysis will be conducted afterwards.

There are two main steps to initial data analysis: (1) cleaning up the data (which itself is split into a screening phase and a diagnosis/editing phase) followed by (2) the preparation and formatting of data for analysis.

4.1.1 Cleaning Up the Data

Cleaning up data consists of assessing the data to check for errors and inconsistencies and then developing solutions to address the errors as well as documenting how you will be handling them.

Before starting to clean up your data, make sure you have backed up the original data set: You will most likely generate multiple different versions of your data throughout the data analysis effort but you must make sure that you always have a copy of the original data. That way, if you need to trace back some original data that has been transformed, you can access the original data to do so.

After backing up the data, the next step of cleaning data is to ensure that it is error-free. This is called the screening phase.[2] You must identify any irregular patterns or inconsistencies, and you must resolve these issues before moving forward. For example, you might notice that one of the data columns is empty, which was due to an error in importing the data from the online survey software to the data analysis software. Other common errors include different variable types in the same columns (mixing numeric and alpha numeric are the most common mistakes) and having a shifted column due to a conversion error.

Once you are satisfied that the database is error-free, you can start familiarizing yourself with the data, identifying more specific issues and correcting them, as well as removing incorrect entries: As a rule of thumb, no more than 5% of the total number of data samples should be rejected.[3]

This is the diagnosis and editing phase of the data and includes the following components:

1. Make sure data from multiple data sources is integrated in a consistent pattern; for example, if there were multiple people doing interviews, you must make sure the data is entered the same way for each interviewer.
2. As for incomplete survey items, there are different approaches. You could look at your minimum sample size; if counting these incomplete survey responses were necessary you to reach your minimum survey size, consider whether you want to (and can) do a new data collection effort to replace them. If it is not possible (e.g., due to time constraints), any incomplete responses included should be cleaned up to avoid any compilation issues. As such, for numerical values, you could decide to replace them with an average value or replace them with a middle

of the range. This will be useful later, if you decide to cross-tabulate them. Another alternative is to leave them empty and adjust analysis accordingly.

3. Check for inconsistencies in the data, such as duplicate records or multiple answers for the same internet protocol (IP) address.
4. Apply any consistency checks that were built into the survey, removing inconsistent entries (see Section 2.3.7 for more information on building consistency checks).
5. Exclude any unneeded data, using the predetermined inclusion or exclusion criteria: For example, if the study was designed to generate a precise image of the U.S. market, remove any non-U.S. responses that you may have collected.
6. Identify the validity of outliers in the quantitative data set. These might invalidate statistical models or create inconsistent results during analysis, so it is important to verify outliers individually. Furthermore, how these outlying values were identified and handled needs to be documented.
7. Scan qualitative answers to your survey to remove entries with unusable data. For example, some participants will answer "awesome" to every single qualitative question, while others will take a moment to complain about the company sponsoring the survey, or the survey itself. As a rule of thumb, I expect that anywhere from 2% to 3% of survey responses will need to be removed due to unusable answers.
8. Double-check group labels and consistency in naming for category variables; while this might sound trivial, it is a time-saver to take the time to format and correctly name data labels, especially during the data analysis phase.

Once these steps are conducted, you should check if the data sample size is lacking, sufficient, or excessive: It is possible that after cleanup, the sample size will no longer be the necessary size to reach statistical significance, and a new phase of data collection will be needed.

4.1.2 Preparation of Data

The final step of cleaning data is the preparation of the data for more advanced analysis. While producing summary tables can be useful, the researcher might decide to adjust the variables for advanced analysis. Some of the data can be used in its unprocessed form but it might be useful to rescale, standardize categories, and transforming some data into a single variable, combining them into summary scores, or using more complex functions such as ratios.[4] As such, you might need to create new variables that will specifically address your research topic—such as creating indices from scales or combining data sets that are too small to be useful categories). Additionally, you must format these variables so they can be processed correctly by the data analysis software. This can include setting a value for missing data codes so that your software can handle them and correctly formatting data variables (so you do not have any numeric and alphanumeric mix-ups).

Demographic data is often the subject of aggregation, specifically variables related to geographic location, age, and educational level. For example, during data collection, a researcher might have collected the state and city of each respondent but might later realize that this level of granularity is not useful for his research objective and decide instead to recode the data according to major geographic regions. It is also possible to recode data if a subgroup is too small to generate useful information. For example, if you did a web survey of 3,000 participants, and less than 1% of participants were aged 65 and older, it might be useful to aggregate them into the closest category and redefine it: So you could combine the data from the 55 to 64 years category with the 65 and more into a single "55 and more" category.

4.1.3 Specific Issues Relating to Cleaning Up Qualitative Data

Qualitative data analysis is more complex than analyzing quantitative data. It is often produced on different medias—including written notes, audio or video recordings, and pictures—which first require transcription into electronic media.

The transcription should be conducted word for word, not just summaries as summaries imply the initial transformation of data: if the transcriber is summarizing data, he is introducing some bias by choosing what information to include, which words to summarize, and so on. Some other things to consider:

- The transcription document should include nominative information, such as who performed the interview, who was interviewed, as well as the time, date, and location the interview took place.
- During transcription, non-verbal cues that were noted by the interviewer should be included in the transcript, as these could prove

informative later on: Mark them distinctively so you can quickly identify them.

- If possible, try to transcribe transcripts yourself—or have someone who attended the interview or focus group transcribe the information, so you can add any thoughts or ideas that emerge during transcription.
- Even if you are trying to be as faithful to the original text as possible, you can edit out verbal tics (e.g., "you know" or "it is like") and fillers (e.g., "uh," "hmm," and more) as these can prove quite distracting during analysis.
- Finally, transcriptions used to be a manual skill, which needed to be conducted "by hand." As we discussed earlier, online tools that you can use to do interviews will now generate transcripts (some will go as far as generating summaries). The time savings are undeniable. But, as of the writing of this book (2026), accuracy remains a key concern. As such, it is important to still take a few notes during interviews; what I will usually do is look at the AI generated transcript and include my notes at the beginning of the transcript to ensure I still have access to them during the analysis phase.

4.2 Data Analysis: Quantitative

Quantitative data refers to data that can be measured and numbered. Counting the number of potential clients for a product, calculating the number of products or doses a consumer uses each day, or the average distance a patient is willing to travel to visit a specialized clinic are all different types of quantitative data.

Quantitative data is easier to share since the results are easy to understand and easy to explain. There is a perception that quantitative data is "hard data" (vs. qualitative data that is perceived as "soft data"). Hence, market research often has a large quantitative component to convince the target audience.

Explaining how to analyze quantitative data is an ambitious endeavor, and we cannot give a complete overview of the process in the limited space we have allocated to it here. There are many fine books dedicated to this topic exclusively, and any reader wishing to learn more could consult *Quantitative Analysis for Management* (12th Edition) by Barry Render (Pearson Publishing) or *The Quantitative Analysis of Social Representations* by Alain Clemence (Taylor & Francis).

This section starts with a presentation on the two different types of analysis: Descriptive and inferential analysis. This is followed by some presentations on univariate and bivariate data analysis models, and some simple tests you can use to verify the significance of the results, such as correlation and regression models, as well as some interesting models that can be used to analyze your data. We conclude this section with a description of computer software packages that are available to analyze quantitative data, and AI will be discussed in Section 4.5.

4.2.1 Overview of Descriptive and Inferential Analysis

There are two types of quantitative analysis: Descriptive and inferential analysis. When doing descriptive analysis, the information describes or summarizes the data that was collected, whereas inferential statistics are used to make inferences from the sample data for a whole population.

Descriptive analysis is used when the researcher is analyzing his data so he can identify patterns in the population he is studying. Contrary to inferential analysis, descriptive analysis will summarize the data in the sample itself, rather than inferring that it applies to a whole population. The advantages of doing descriptive analysis are multiple: First, doing descriptive analysis makes vast amounts of data both more manageable and organized. It is also straightforward, can be used to further research ideas, and lays the groundwork for more complex analysis.

In this book, we will be looking at descriptive analysis from a univariate analysis angle (with a focus on frequencies of variables, central tendencies, and dispersion) and a multivariate analysis angle (with emphasize on the relationships between variables).

The restriction of descriptive analysis is that it is limited to the data you are handling. It is used to report and describe data that you collected but cannot be used beyond the current data. The researcher cannot use the data to interpret phenomena beyond the population from which the sample was taken. For example, if you tested how a product is used by a sample population, you could not consider the findings representative of the population at large using exclusively descriptive analysis.

Inferential statistics would be needed. Therefore, while descriptive statistics are used to describe what is happening in the sample population alone, inferential statistics are used to infer what the population beyond the data sample is thinking. Hence, inferential statistics are techniques we use with our data set to generalize about the populations from which the samples were collected.

The limitations are that inferential analysis is uncertainty. While calculating the proportion of a sample in a specific situation is absolute, projecting it onto a population means that it becomes an estimate: Building this estimate requires the researcher to make educated guesses, and this uncertainty will increase the likelihood of errors.

There are several techniques that market researchers use to examine the relationships between variables, thereby creating inferential statistics. We will look at some of the most valuable ones in market research: Correlation, regression analyses, and general linear models.

4.2.2 Univariate Analysis

Univariate analysis is the simplest way to analyze data. All that it entails is the analysis of a single variable, which can be categorical or numerical. Since it is not being compared to one or more other variables, it is very descriptive in nature. Univariate analysis cannot be used to explain a phenomenon or to contextualize a variable in relation to other variables. Once data is compiled, it can be shared in a simple table form, as well as in visual representations such as bar charts, histograms, frequency polygons, or pie charts.

The most frequent univariate analysis methods are frequencies of variables, central tendencies, and dispersion.

4.2.2.1 Frequency Distribution, Central Tendency, and Dispersion

Frequency distribution is a simple data analysis technique that allows the researcher to get a quick impression of the data he has collected. Using frequency distribution, he can see how often the specific values are observed and their percentages compared to the overall sample. As such, frequency distribution enables the researcher to summarize each variable independently and makes it easier to subsequently engage in more complex analysis.

To illustrate frequency distributions, we will use data from a project I did a few years back. It was for a company in the dental health space. This client had developed an innovative dental product and was looking to evaluate how much interest dentists would have in the product, according to various distribution models. Over 100 dentists (current and past clients) participated in an online survey. The survey included both quantitative and qualitative questions.

To validate interest in the innovation, the client had asked the following question: "How likely would you be to use [product description] if you had easy access to it and the turnaround times described above? (i.e., less than 90 minutes)" The data was compiled and illustrated in both a table (which is more detailed and descriptive) and a graph (a more visual format to share data) (Table 4.1; Figure 4.1).

Based on this univariate analysis, there seemed to be some strong interest in the current client base, as 86.0% of participants were either "somewhat likely," "likely," or "very likely" to use the product developed by my client.

4.2.2.2 Central Tendency

There are many ways to find the central tendency of a data sample. You can use either the mean (the sum of the values divided by the number of effective values), the mode (the most frequently occurring value), or the median (the value separating the upper half of a data sample). This is the traditional method of presenting scoring data and giving a sense of the

Table 4.1 Frequency Distribution Example of Survey Data

Research question: "How likely would you be to use [product description] if you had easy access to it and the turnaround times described above? (i.e., less than 90 minutes)"

	Frequency	*Relative %*	*Valid %*	*Cumulative %*
Very likely	32	28.8	29.9	29.9
Likely	26	23.4	24.3	54.2
Somewhat likely	34	30.6	31.8	86.0
Somewhat unlikely	6	5.4	5.6	91.6
Not likely	9	8.1	8.4	100.0
Total	107	96.4	100.0	
No answers	4	3.6		
Total participants	111	100.0		

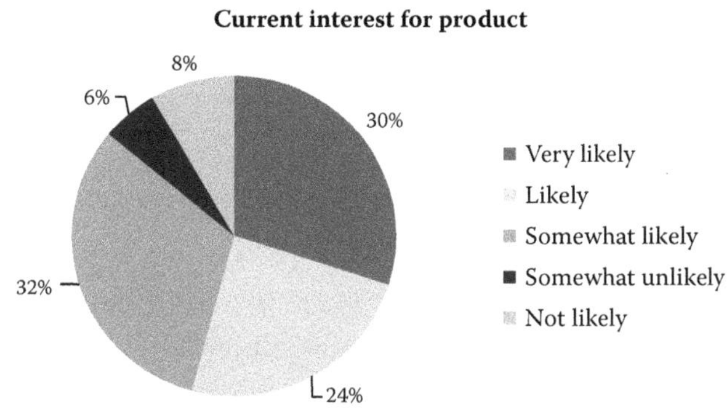

Figure 4.1 Frequency distribution example of survey data in a classic pie chart.

general appreciation in a population (e.g., statements like: "Participants in our survey gave an average of 7 out of 10 for the overall look of our new product label").

Using a mean is the most popular way to calculate central tendency, as it includes all the values of your data sets, but it is particularly sensitive to the presence of outliers. The outlier can skew the mean, rendering the data unusable. Another reason for not using the mean is if your data itself is skewed: If, for example, half your sample would be willing to pay $10 for your product, and the other half would be willing to pay $20, a mean of $15 would not reveal that you probably have two distinct market segments in your data rather than a single monolithic segment.

As for modes, these demonstrate the most popular option in the sample. The problem when using the mode is when it is not unique: If you have two data sets with the highest frequency, it can be challenging to interpret the data. Another challenge with the mode is that it does not give a good read of the central tendency: If you asked each participant to give a score of 1–10 on the quickness of a product, and then half responded 1 out of 10, the mode would not be representative of the central tendency.

The median is not affected by the presence of outliers, so it gives a better idea of a typical value. It is also better to use when the distribution of the sample is skewed. The main issues when using the median (rather than the mean) are that it is less popular than using the mean to communicate the central tendency, so people might not be familiar with its meaning. Also, a medium is much more complex to calculate if you are not using software, which honestly should not happen too often if you are preparing a market report.

4.2.2.3 Dispersion

Dispersion is the calculation of the distribution of values around a central value. The three main dispersions that you can calculate are range (the distance separating the highest and lowest value), the variance (the average of the squared differences from the mean), and the standard variation (which is the square root of the variance and measures how much the numbers are spread out).

As an example, a project I worked on surveyed over 5,000 individuals on their internet usage. One of the questions asked to participants was to estimate time spent online. Across all age groups, the average was 19 hours, with a standard variation of 12 hours. This meant that most of the data sample had answered between 7 and 31 hours spent online, giving us a rough idea of the time spent online for the surveyed population.

Calculating dispersion is important to understand the spread around the mean, or the variation around central data. The mean of a data set indicates the average, but the dispersion will supply you with invaluable information on how much of your data sample is grouped within a certain distance of your mean. The lower your dispersion (Figure 4.2), the more your mean is representative of your central tendency. The higher the dispersion (Figure 4.3), the more dispersed your data set is.

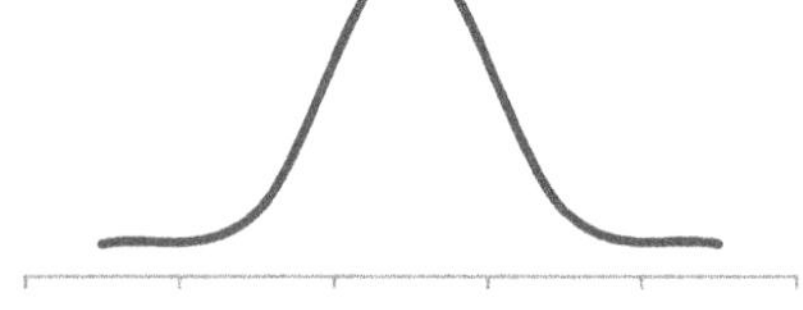

Figure 4.2 Representation of low dispersion.

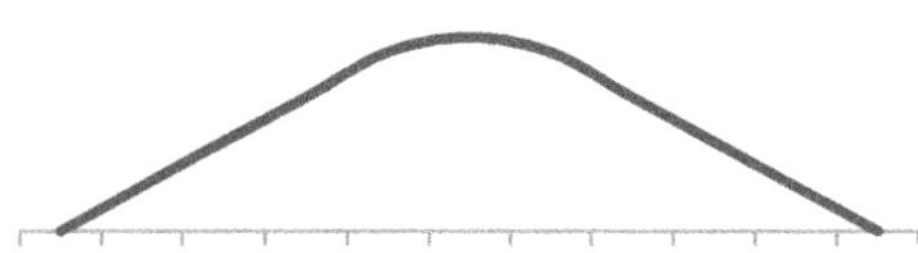

Figure 4.3 Representation of high dispersion.

4.2.3 Multivariate Analysis

Multivariate analysis is the concurrent analysis of two or more data sets. While univariate data analysis is used to *describe* a phenomenon, multivariate analysis is used to *explain* the phenomenon.

The objective is to compare two (or more groups), and to identify relationships between the multiple data sets. The analysis exposes how different subgroups respond to some query. The simplest method of doing multivariate analysis is cross-tabulation (or contingency tables), while more complex tools include correlation and regression analysis.

4.2.3.1 Contingency Tables

A simple way to do multivariate analysis is to build contingency tables, which are multidimensional frequency distributions in which the frequencies of two (or more) variables are cross-tabulated. These simple tables can help the researcher find the relations among different variables. Contingency tables are called pivot tables in Excel.

To properly illustrate a contingency table, we are sharing a table illustrating bivariate analysis (Table 4.2).[5] The data for this table was for a project I did a few years back, the theme of which focused on "what women want." Surveying Australian mothers, one of the topics the survey dealt with was identifying how mothers got information relating to their pregnancy. This specific table cross-tabulated data from two questions:

a. Did you use this information source to get more information relating to your pregnancy? (Check all that apply.)
b. What is your age?

Immediately, there are four apparent key learnings we get by doing a cross-tabulation (Table 4.2):

1. It seems that women younger than 39 years old are much more likely to use online resources to look for information than their older counterparts.
2. It seems that women younger than 39 years old are much more likely to consult their mothers for advice than their older counterparts.
3. Only half of the participants contacted the Department of Health to get information relating to their pregnancy.
4. There is an overwhelming use of pregnancy books and manuals across all age groups.

Table 4.2 Example of a Bivariate Contingency Table: How Women Find Information Relating to Their Pregnancy versus Age of the Respondent

		Age		
		18–24	*25–39*	*Over 40*
Criteria	*Used*			
General online resources	Yes	**81**	**800**	72
	No	38	477	151
Department of Health	Yes	40	421	59
	No	79	856	164
My doctor	Yes	79	838	144
	No	40	439	79
My friends	Yes	71	797	142
	No	48	480	81
My mother	Yes	86	762	92
	No	33	515	134
Other family members	Yes	56	541	77
	No	63	736	146
Pregnancy books and manuals	Yes	88	1120	196
	No	31	157	27
Pregnancy-focused websites and email newsletters	Yes	**87**	**896**	94
	No	32	381	129
Pregnancy forums and communities	Yes	**62**	**504**	39
	No	57	773	184

Overall, a contingency table is an effective and simple way to find underlying information that univariate analysis would not have been able to discern by using the data from two distinct data sets.

4.2.3.2 Correlation

Correlations are an extension of a cross-tabulation analysis. While building a contingency table might uncover a potential relationship between multiple data sets, correlation analysis is necessary to confirm the relationship between the variables. To confirm the existence of a correlation between two variables (as well as regression and general linear relationships), the use of specialized software such as SPSS or JMP is recommended, although it is possible to use Excel as well.

When calculating correlation, for each data set, the software will provide an answer ranging between 1 and –1. The closer the answer provided is to 1, the stronger the relationship between the two variables. The closer it is to zero, the more it means that there is no relationship between the two variables. If you obtain a correlation of –1, this means that your two data sets are negatively correlated, meaning the higher that one value goes, the lower the second one goes. As a rule of thumb, if you obtain a correlation value of 0.5 or higher, this means that there is a significant relationship. This becomes interesting as you start to calculate the strength between multiple sets, and you can statistically see which relationships are stronger and determine key elements, or those that are important for the participants in your research.

There are two major caveats to calculating correlation. First, *correlation is not causation*. This means that you might have found that two data sets have a relationship, but there is no proof that one element is causing the other. There could be multiple reasons why your data is correlated. Second, calculating correlation is only possible if your data is in a linear relationship. A good alternative if you believe that there is a relationship between two data sets, but that your calculations or correlations do not support it, is to do a scatter plot of your data so you might be able to visualize a non-linear relationship.

4.2.3.3 Regression Analysis

Regression analysis is another way to estimate the relationship between two variables. It is a test used to see how one independent variable affects the other dependent ones. Some prefer it to correlation since it statistically demonstrates the goodness of fit (adjusted R Square). Regression analysis can also be used for forecasting, using the regression equation as a simple linear model.

One of the main differences between correlation analysis and a regression analysis is that while correlation analysis quantifies the relationship between variables, a regression analysis can be used to predict the relationship between variables, especially if one of the two variables is one that you manipulate (e.g., time or price).

Table 4.3 Guidelines for Interpreting Correlation Coefficients

If the coefficient is . . .	*Then the correlation is . . .*
Between 1 and 0.7 then	Strong
Between 0.7 and 0.4 then	Moderate
Between 0.4 and 0.2 then	Weak
Less than 0.2	No correlation

As a rule of thumb, the guidelines for the interpretation of your regression factors are as shown in Table 4.3.

4.2.3.4 General Linear Model

General linear models are used to assess the effect of predictors on one or more continuous variables. They include tests such as the *t*-test (which is used to examine the difference between the means of two independent groups) and the ANOVA test (tests the significance of differences between the means of two or more groups). These are useful to assess the effect of several predictors on one or more continuous dependent variables.

4.2.4 Software for Quantitative Analysis

There are two types of software that you can use for analyzing quantitative data.

First, it is possible to do a lot of basic and even some intricate analyses using spreadsheet software such as Excel. Spreadsheets can calculate anything from simple averages (with the AVG and MED functions) and frequencies (using the FREQUENCY function) to more complex calculations such as standard deviations (STDEV), correlations (using the CORREL function), or even trends (using the TREND function, calculating regressions). There are a number of resources online on how to use advanced functions, but one of the more interesting books I have used is *Excel Statistics: A Quick Guide,* written by Neil Salkind and published by Sage Publications.

The main advantage of using a program like Excel is that many users are already familiar with spreadsheets, so the learning curve to using it is smooth. Also, integrating data from other software (e.g., online websites used to collect survey data) is straightforward. Finally, as you are already formatting your data in your spreadsheet, preparing your data for presentation models (tables or graphs) is streamlined. As such, spreadsheets are a good alternative for companies that do not engage in statistical analysis on a regular basis.

For more complex analysis, or if you engage in quantitative analysis on a regular basis, obtaining a statistics software package is a necessity. There are several of them available, but here are a few of the more popular options:

- One of the market leaders is IBM SPSS Statistics (www.spss.com). It is a powerful statistics software package used for statistical analysis in academic and business environments. It is quite popular with scientists and marketers alike and contains a full suite of tools that are useful for any researcher. It has a wide range of charts and graphs to choose from and efficient access to statistical tests. While powerful, it is one of the costlier solutions out there.
- SAS (www.sas.com) is another prevalent analytical software program that is used for advanced analytics and analysis. It is positioned as one of the most powerful software programs available but is quite complex to program and utilize. It is one of the hardest software programs to learn but offers great data management and the ability to work with multiple files simultaneously.
- A less expensive option, albeit still powerful, is JMP (www.jmp.com). Like SPSS, JMP has several options for data analysis and produces a full range of charts and graphs. It is very straightforward to use and somewhat less expensive, and easily imports data from most formats, including SPSS tables.

There are also several free statistics software packages available but be aware that the learning curve is steeper than some of the commercial packages available. One of the popular free alternatives is PSPP (http://www.gnu.org/software/pspp/pspp.html), which, while not as powerful as the commercial alternatives, does have a similar look "n" feel to SPSS, which makes the transition easier. It also has an easy point' n click interface.

Final note: The use of these different specialized statistical software continues to be present, especially in pharmaceutical and biotechnology as they leave a trail that can be audited and validated. However, for simple analysis, AI agents are slowly replacing these tools.

4.3 Qualitative Data Analysis

Qualitative data is information that is subjective and subject to interpretation. It can include anything from stories to words, observations, pictures, and even more peculiar sources such as songs, stories, and poems. Examples of information you can collect through qualitative analysis include personal preferences in consumer purchases, the impact of quality on customer purchasing decisions, or the influence of packaging color on acquisition decisions. Data collected through interviews, focus groups, Delphi groups, and observation is usually qualitative in nature, but as we will see, it is possible to codify the collected data so it can be analyzed using quantitative data tools.

Qualitative research is particularly useful when the objective is to find out what people are really thinking, identifying future trends, and speculating on competitive threats. Strategic thinking and careful analysis are needed to identify those trends. In the next few pages, we will go over a proposed framework for qualitative data analysis, followed by a discussion of computational tools available to enhance qualitative data analysis.

4.3.1 Qualitative Data Analysis Process

Quantitative data has the advantage of being simple to analyze. It is often collected in a format that comes pre-coded and ready to analyze, whereas qualitative data needs to first be codified and labeled before proceeding with the analysis. To be able to code qualitative data, the researcher will need to build a classification framework, following a process to categorize verbal or behavioral data to enable the classification, summarization, and tabulation of his data.

The four steps of the basic qualitative data analysis process are (1) familiarization, (2) identifying a framework, (3) sorting the data into the framework, and (4) using the framework to complete the descriptive analysis.

Some researchers suggest doing an early analysis step before diving into the complete qualitative data analysis. This implies analyzing some qualitative data during data collection. They believe that it is an effective method to optimize ongoing research and generate new ideas for collecting better data. It can also give you an early idea of what your analysis framework will look like. I believe this should be avoided in studies that are exploratory in nature, to ensure that the first few interviews do not direct you down a path that has not yet been validated. For example, if you do an interview with a few doctors to explore the important factors for signing up for your new service, and they focus on the login and payment screen, there is no indication that these concerns are valid throughout your population, and focusing on them too early in the study might bias the final outcome.

4.3.1.1 Step One: Familiarization

The first step of qualitative data analysis is to familiarize yourself with your data and what it looks like and then start to visualize the necessary effort for data analysis. This means re-listening to multiple interviews, reading multiple interview transcripts, and reviewing open-ended survey answers as well as rereading the notes you collected during the research effort. If you worked as part of a team, a debriefing meeting will be useful to discuss high-level impressions as well but be careful that these discussions don't bias the next analysis steps. At this stage, you are trying to identify the most relevant and interesting observations from your data.

4.3.1.2 Step Two: Identifying a Framework

The qualitative data framework is the coding plan that you will be using to organize your information. During this step, you will be identifying high-level data patterns that you will use to build your first framework.

This framework will consist of *codes,* which are tags or labels that are used to assign meaning to qualitative information. They are usually attached to pieces of text, such as words, expressions, phrases, and even paragraphs. They can take the form of a simple word, or can be more complex, in the form of expressions and metaphors. These codes are used to create meaning through clustering: The more often a code appears during analysis, the more meaningful the meaning behind it. These code clusters are then used to analyze data, identify patterns, and generate recommendations or concepts.

When developing your coding, it is possible to use three types of codes:

1. *Description codes:* Codes that describe the phenomena and require very little interpretation.
2. *Interpretative codes:* When you have a bit more knowledge about what you are researching, you can start assigning more complex coding by using codes that integrate some interpretative elements.

3. *Pattern codes:* These are codes that you can start using when some of the more complex elements such as patterns, themes, and causal links are identified in more depth. This might require you to go over some earlier analysis to verify if some codes need to be updated.

There are multiple methodologies that can be used to build a framework and start identifying your codes, but we will be going over three of the more common ones. These are manual iteration, automated text analytics, and directed content analysis.

Method 1: Manual Iteration

Manual iteration is the most complex and lengthy method to develop a framework but it is the one that will less likely result in you having to go over your data twice.

The first step is to read through a few records to get a high-level appreciation of which phenomena are occurring most often. If the bulk of your qualitative data is interviews, review a few transcripts and start identifying key trends. If it is focus groups, read through the transcripts and moderator notes to identify those common elements. If it is a survey with qualitative open-ended questions, read the first 100 responses, and identify keywords, ideas, and thoughts in a separate column of the data analysis tool that you are using. Afterwards, go through the new column of keywords that you generated to identify patterns and build a framework for analysis. If a consistent framework cannot be found, read a few more interviews, or a second group of 100 survey responses, and then look through your notes to identify that pattern.

To illustrate the process in an online survey, I have included a short extract from a project I analyzed a few years back.

To give some context, my client, a company active in event organization and media platforms targeting the Generation Y public, had organized a survey targeting its subscribers. One of the issues they investigated was the purchasing patterns for healthy drinks. Hence, they had asked participants to think about the last drink they had purchased, and the reason that had led to that purchase over another competing brand or product. My client had collated the data and brought me in to analyze it.

Hence, the first step was to build an analysis framework. I chose to do it through manual iteration since the sample was quite large and complex, and the context was exploratory. To demonstrate, I have included a few statements taken from the survey, followed by some tags that I felt were suitably related (Table 4.4).

Table 4.4 Example of Building a Codification Framework through Manual Iteration

Statement	*Keywords*
A friend recommended it to me; next time I saw it in a shop I tried it, and he was totally right.	Friend's recommendation, convenience
Because they were the cheapest!	Pricing
I bought it because of brand loyalty, design, hype, and health.	Brand loyalty, design, popularity, and health
I am trying to get healthy, and I had heard about it through family.	Health, family recommendation
Word of mouth from friends that the product is superior to others	Word of mouth, friend, recommendation
Recommended reviews online, a well-known/trusted brand around for a long time, good price.	Online review, brand reputation, trust, price
I bought it because it was on sale and mostly because I do not want to fill my body with industrial waste anymore.	Price, sale, health, ethical

Other patterns were identified around themes such as "locally made," "great website," "celebrity endorsement," "advertising," and more. Not all these keywords were kept in the final version of the framework: Some keywords, which were present in the first few entries, had very little presence in the survey responses overall, and were discarded. Hence, because the process is iterative, some keywords that are identified in the first step might be dropped later on, while others are added during the more in-depth analysis. Some overlap might also occur, as well as splitting up some keywords of your framework.

You might note that there is a high level of subjectivity in this process. While I use "friend's recommendation" to denote the phenomenon of "a purchase due to a recommendation of a non-family-related person," you might prefer using the code "third-party recommendation" to widen the definition to "a purchase due to a recommendation of a third party." This is all part of the strength and weakness of qualitative data. Future iterations will enable you to refine your framework—you might find that there is enough of a distinction between friends and parents to warrant a second category. Or you may not. Flexibility is the key at this stage.

This is also the reason why two researchers need to work in tandem on some of the bigger projects. By going over the work independently, and then combining the results afterwards, the researchers can develop some level of consistency.

The framework that you build will allow quicker analysis of the responses that you have collected. By the end of this step, you will have identified a series of keywords and will be ready to advance to step three.

Method 2: Automated Text Analytics

This method can be used with the qualitative analysis of data sets for which the answers are short yet in high quantity, such as qualitative answers in an online survey. The idea is to use a word cloud generator.

Word cloud generators are platforms that count the number of times each word appears in a data sample, assign it a relative density, and then generate a visual representation according to the word density of the top words. There are many free word cloud generators that exist online such as WordItOut (worditout.com/word-cloud/create) and WordCloud (www.jasondavies.com/wordcloud), which you can use to generate your own word cloud for both analysis and presentation purposes.

After choosing your word cloud generator, copy a random number of entries for the topic you are analyzing into the text box. The platform will generate a word cloud, giving more prominence to the words that appear more frequently. These words can be used as a basis for the first framework for data analysis. Be advised that while this method is much easier and faster than manual iteration, this shortcut will likely overlook some keywords, and some revision of your framework might be needed later.

To illustrate, using the same data set as used in the first method we generated a word cloud using 100 random answers from participants. The example was generated using the WordCloud website. Remember to exclude common words to increase the visual effect of the word cloud. In this case, we excluded words such as "because," "around," and other commons articles that are generally not important (Figure 4.4).

A quick review of the word cloud lets us appreciate the emphasis that participants place on the quality of the product, such as the brand name, the reputation, recommendations, and reviews, for example. Using the word cloud, we can start the next step of analysis with a half-dozen or more code words, such as quality, price, reviews, friend's recommendation, brand, taste, trust, and healthy.

Figure 4.4 Word clouds generated by online word cloud generator "WordCloud."

Method 3: Directed Content Analysis

Finally, some researchers use directed content analysis. Using this methodology, some of the codes are developed before data analysis, using theory and a review of the literature or secondary research to build the question guide and the theoretical framework to guide analysis. Using this method, additional codes are added as the analysis progresses. The advantage of this method is that the existing framework lets the researcher jump into the analysis phase faster, and the questions guide can be built according to the developed analysis framework. Nonetheless, this is less useful in exploratory contexts and can bias some of the data collection as well as the analysis.

Coming back to our example presented earlier in this chapter, if we had decided to use directed content analysis for our project, there are several ways we could have built our framework:

- If we had built it on the client's experience (hence a "hypothetical framework"), the client and I would most probably have used terms such as *cool, price,* and *friend's recommendation* as starting points for our framework. (Based on a quick review of our email conversations, these are the terms he expected to emerge.)
- Alternatively, I could have reviewed an article such as "Marketing to the Generations"[6] from the *Journal of Behavioral Studies in Business* and identified "prestige," "uniqueness," "pricing," and "referrals" as starting points.

In both cases, the results are the same: A pre-existing framework is imposed over qualitative data. Even if there is a speed advantage, there is a potential negative impact on the analysis of exploratory data that should not be underscored.

4.3.1.3 Step Three: Coding the Data Using the Framework

Once a suitable framework has been built, it is time to assign the codes to your data. As this is an iterative process, expect to modify your framework during analysis: You will most likely remove, merge, and add codes all the way during your analysis process. This might mean that some data elements must be reviewed multiple times. This is a normal step of qualitative data analysis. In this section, we will illustrate two examples of coding qualitative data, one for interview data and one for survey data.

To code interview data, I usually use a three-column analysis worksheet. The first column has coding that is attributed to a sentence. The second column is the interview itself, with phrase segments allocated to phenomena, and the third column is general comments and thoughts that emerge during analysis. These thoughts are often very useful during the iterative process, to help adjust the coding categories, to help define them precisely, and to prepare for the next step, pattern coding.

To illustrate this process, here is a short extract analysis of an interview I did a few years ago. It was for a company planning to offer outsourcing services to large companies and needed to better understand the trends impacting the need for specialized outsourcing abilities. The company needed to understand decision-making processes and identify growth opportunities. A review of the literature had enabled us to build the framework shown in Table 4.5 beforehand.

The codification then enables quick clustering and analysis: Instead of looking through hundreds of pages of interviews, you can quickly scan

Table 4.5 Sample Coding Framework

Code	*Definition*
REL	Relationships
OUT.GO	Outsourcing to global players
OUT.LO	Outsourcing to local players
VIR	Virtualization

Table 4.6 Example of Analyzing an Interview Extract

Codes	*Text*	*Notes*
OUT.GO OUT.LO VIR REL	JFD: One of the key issues I wanted to explore is if there are any preferences for local providers versus international providers. MM: At our company, we prefer global providers to match our global operations, but then we prefer to use global providers that have a presence near our strategic offices. JFD: Why is that? MM: We have found that virtualization is a thing, but sometimes problems require a face to face. Also, it is just better to be able to interact directly JFD: Has that shifted the nature of the outsourcing you do? MM: It has. It means looking for the right partner. I believe organizations will shift strategy, taking a longer-term view of sourcing options instead of a one-time decision on which functions to outsource and those to keep in-house. Provider selection is changing. It will take greater effort to cut the field to four finalists who receive the request for proposal (RFP).	N1: Looking for global players that are local, best of both worlds . . . N2: Seems to be a recurring theme, focus on the long-term relationships

and look for patterns or codes that reappear often and then group these comments to craft your story. Codes make the interview more intelligible, and the recurrence of certain codes signals the emergence of regular themes (Table 4.6).

Survey analysis follows a similar approach. Following up on the survey on purchasing patterns for healthy drinks, we coded all the responses from the participants. Each statement was individually analyzed and assigned a code from our framework. As the analysis proceeds, new codes can emerge, and new ones can be created from consolidation. For example, during the consolidation stage, the categories "user reviews" and "online reviews (blog)" was consolidated into user reviews, as the number of responses did not justify keeping those two categories separated.

Table 4.7 Example of a Matrix of Survey Data Analyzed

	Specifications and Features	*Brand Loyalty*	*Personal Favorite/I Wanted It*	*Great Quality*	*Family Re-commendation*	*Pricing*	*Availability/ Convenience*	*Looks Great*	*Friend Re-commendation*	*Healthy*
A friend recommended it to me, next time I saw it in a shop, I tried it, and he was totally right		x					x		x	
Because they were the cheapest!						x				
I bought it because of brand loyalty, design, hype, and health	x	x						x		x
I am trying to get healthy and I had heard about it through family					x					x
Word of mouth from friends that the product is superior to others				x					x	

Table 4.7 ***(Continued)*** **Example of a Matrix of Survey Data Analyzed**

	Specifications and Features	*Brand Loyalty*	*Personal Favorite/I Wanted It*	*Great Quality*	*Family Re-commendation*	*Pricing*	*Availability/ Convenience*	*Looks Great*	*Friend Re-commendation*	*Healthy*
Recommended reviews online, a well-known/trusted brand around for a long time at a good price		x				x				
I bought it because it was on sale and mostly because I do not want to fill my body with industrial waste anymore						x				x

To ease codification, we built an analysis matrix. Qualitative data was stacked on the lines, as columns were each assigned a framework item. As the process was iterative, it was possible to add codes as new elements were found. To best illustrate this, I have included a sample extract of the matrix that was generated (Table 4.7).

4.3.1.4 Step Four: Use the Framework for Descriptive Analysis

Once data is analyzed, the codes are used to create significance using clustering. The more often that codes appear during analysis, the more expressive the concept behind them. These clusters are then used to identify patterns and generate recommendations.

For example, in Table 4.7, we see clusters emerge around the concepts of brand loyalty, pricing, and health. That is one of the advantages of using a matrix data analysis, as it provides a medium for an initial visual interpretation of data. The second advantage is that the data can then be converted into quantitative data. By compiling the tags, we can create a table, and generate the data as shown in Table 4.8.

As for interviews, the next step is pattern coding. Here, the objective is to group the categories into smaller sets of themes and constructs. Hence, during this step, recurring phenomena emerge, and you can see

Table 4.8 Example of Compiled Qualitative Data: Top Ten Reasons Readers Purchased Their Last Drink over Another Competing Brand or Product

Reason	*N*	*Percentage*
Pricing	234	23.4
Brand reputation	166	16.6
Great quality	150	15.0
Friend recommendation	97	9.7
Specifications and features	88	8.8
I needed it	79	7.9
User reviews	78	7.8
Familiarity	77	7.7
Personal favorite/I wanted it	71	7.1
Advertising	61	6.1

the most important ones. Finally, remember that it is very important to carefully define each theme; if they are not properly defined, some blurred lines might complicate the analysis, and render the framework invalid, or ill-defined.

4.3.2 Some Final Notes about Qualitative Analysis

As we have mentioned numerous times throughout this book, qualitative data is often alleged to be less "real" than quantitative data. This might be due to the ease with which it can be manipulated, or how the data sample is often so small relative to large quantitative studies that it is dismissed as anecdotal. This is simply not true, and qualitative researchers can follow several practices to refute this perception, such as using a rigorous qualitative analysis framework, using computational tools to assist qualitative analysis, and using mixed methods of data analysis.

Also, as qualitative data is more subjective in nature, it is often useful to work with multiple analysts when analyzing it (depending on the project scope, of course). If the budget allows it, two (or more) different analysts should analyze the same data following the same methodology, which would be followed by a convergence step and discussion on divergent results and interpretation.

4.4 Creativity in Data Analysis

Data analysis requires both a systemic mind as well as some creativity. As you look at your data, you might find yourself struggling to identify patterns or solutions to problems your users have shared with you. A book I wholeheartedly recommend here is "A Whack on the Side of the Head," a classic creativity book by Roger von Oech, which offers puzzles and tips to overcome mental blocks. It should be a part of every marketer's library. Here are a few of my favorite techniques.

- *Look for the second right answer:* Sometimes data will suggest an easy and obvious answer. And you might think to yourself, everybody must be coming to the same conclusion. So how can I differentiate myself? This is where looking for the second right answer comes in. In this technique, you deliberately put aside the "easy" answer and start thinking of a second answer that also fits your data set. For example, your data might have found the earth-shattering fact that consumers prefer lower prices. Your first instinct might be to lower prices. But what if you took it one step further? They might prefer lower prices because they purchase this often and hence are more sensible to bulk purchases. They might also see this as a commodity good with little differentiating factors. In both cases, these would inform your marketing strategy and might warrant more data analysis.

- *What if?* In this technique, ask yourself "What if?" with a contrarian condition or fact, then try to answer this question. For example, imagine you are analyzing a customer survey for a rapid diagnostic device, where users answered speed was their main interest. You might be tempted to analyze this as "speed is a key attribute." But Rapid tests are, by their nature, rapid. Hence, you have not found any unique attributes. So, you might ask yourself, well "What if my Rapid diagnostic was slower than the competition?" You could decide to develop a safety narrative "Built-in verification time catches what rushed tests miss," or the emotional angle "Reduce the panic of instant results" or even the contrarian approach "In a world obsessed with fast, we are obsessed with right."
- *Create an oracle of creativity:* This is a technique that might require the participation of multiple individuals. First step, address a question to the "oracle." Then generate a random piece of information. Last step, interpret the random piece of information as the answer to our question.

  As an example, imagine you had conducted a web survey, and found that over 70% of participants preferred the bulkiest version of your medical device. So, my question to the oracle could have been "Why do consumer prefer the bulkier version of our device?" Using Claude to generate a random answer, I got is "The Pacific Ocean covers more area than all land masses combined." Some of the creative interpretation could be that people already have a vision that our device will be bulky anyway, people are used to bulky devices, or the people trust devices that are substantial, as it signals capability, reliability, and completeness. Sometimes market preference is not about features or benefits, it is about matching the archetype that already exists in people's heads. The next step would be to test the market's perception and see their flexibility to smaller devices.

  All in all, these three creative techniques (and many more) could help you go beyond a more surface analysis.

4.5 Using AI to Assist in Data Analysis

AI greatly enhances a researcher's ability to process and analyze large amounts of data. It can ease the burden of manual transcription and reduce miscalculations due to human error and facilitate researcher tasks.

As we move forward, there is a definitive trend toward using AI for data analysis. Speed, accuracy, and sexy visualization tools, all within reach with a couple of clicks, will change the relationship between the researcher and his data. Furthermore, as AI refinements increases, we will see an improvement in basic data analytics steps such as data clean up and even predictive analysis.

Nonetheless, these tools should not replace the researchers' ability to interpret and read his data. The use of these tools will not teach the user basic statistics, and without these basic skills, it becomes very easy to believe the data "because my computer says so." The ability to interpret data is crucial going forward, even if tools evolve in directions that make this easier. must realize that while using AI will make the codifying and analysis process much easier, it also distances the researcher from his data, and he loses some of the opportunity to understand his data.

When using AI for data analysis, it is important to understand that there are two distinct approaches: Analyzing your own data that you provide to the AI or asking the AI to gather and analyze information on its own. The latter approach requires caution and critical thinking. When AI collects its own data through web searches, the results can appear authoritative and comprehensive, but the underlying sources may tell a different story.

For example, when I asked an AI agent "how do train companies perceive the use of biolubricants?" It provided a confident, definitive answer: "Train companies are increasingly adopting biodegradable lubricants, also known as biolubricants, due to their environmental benefits and operational advantages." However, upon examining the sources the AI actually used to reach this conclusion, the foundation was quite limited. The sources included AW Solution (a company that produces train lubricants), NEOL Copper Technologies (a company that produces eco-friendly train lubricants), Kluber Lubrification (a specialty lubricant manufacturer), Nextant.ECA (an advisor for energy, gas, refining, and chemical industries), and a market study from Fortune Business Insights. In other words, the AI's sweeping conclusion about industry-wide perception was based primarily on marketing materials from lubricant manufacturers and industry consultants, which were hardly representative of actual train companies' views. This highlights why it is crucial to verify the sources behind AI-generated analyses rather than accepting their conclusions at face value. Overall, there an inherent risk when the AI is both the researcher *and* the analyst. It is choosing what to look at, what to prioritize, and how to synthesize it, all without transparency about those choices. Also, AI also cannot easily distinguish between a press release and genuine industry sentiment, or between widely held views and niche perspectives that happen to be well-represented online.

When working with AI for data analysis, it is generally more reliable to have the AI synthesized information that you have already selected and vetted yourself. For example, if you provide it with 10 interviews you have conducted, 20 scientific articles you have chosen, and statistical data from government sources that you share directly, you have already conducted the critical work of curating quality sources. The AI then acts as a synthesis tool rather than a researcher, which significantly reduces the risk of it drawing conclusions from inappropriate or biased sources like manufacturer marketing materials.

However, even this approach has important limitations you need to be aware of. First, you should never share proprietary or confidential information with certain public AI models, as they might use it for training purposes or potentially to answer other users' questions. Second, the classic "garbage in, garbage out" principle applies—if your curated selection is biased (for instance, if you only chose articles supporting your hypothesis, or only interviewed people likely to agree with your position), the AI will simply reinforce that bias in its synthesis without flagging the problem. The AI will not tell you that you have created an echo chamber. Third, most AI models do not perform any weighting or quality assessment of the information you provide—the opinion of a leading expert in a field and the opinion of an undergraduate student carry the same weight in the AI's analysis. Similarly, AI does not evaluate the quality or freshness of your sources on its own; a 2025 study and a 2005 study will be treated as equally current and relevant unless you specifically instruct otherwise. This means how you frame your query and provide context is key, and you need to explicitly tell the AI things like "prioritize the more recent studies" or "weight the expert interviews more heavily than the student perspectives" if you want it to handle the information appropriately.

It is also important to understand some broader limitations of AI models when synthesizing information. First, AI systems vary significantly in how well they handle conflicting information in your sources. Some will try to smooth over contradictions and present a falsely unified narrative, while ideally, you would want them to explicitly flag where sources disagree and explain the different perspectives so you can make informed judgments. Second, AI models often struggle to conceptualize abstract business concepts that experienced practitioners understand intuitively—things like bulk pricing dynamics, bundling strategies, bargaining power relationships, or early adopter adoption processes. The AI might use the terminology correctly but miss the nuanced implications that would be obvious to someone with domain expertise. Finally, AI have no common sense. For example, an AI could easily equate people waking up and eating breakfast to indicate that waking up makes you hungry.

This is why the end user's expertise is an essential component of reviewing AI-generated analysis and identifying hallucinations. One found a post on an Ars Technica article where the user mentioned "In short, everything AI spits out is hallucinations, it is just that most of them are close enough to reality so we do not know." While technically true (AI does not "know" anything; its generating responses based on statistical probability), there are some data points that are so well reinforced by training data that the risk of error is exceedingly small. AI will not ever mistake London for the capital of France, but it might have challenges finding you the best place to eat good Mauritian dholl puri.[7] If you personally conducted the interviews and read the documentation you are feeding the AI, you are well equipped to spot when the synthesis goes off the rails—when it misinterprets a quote, invents a connection that does not exist, or fundamentally misunderstands a business dynamic. However, if you are blindly trusting the AI's output because someone else did the interviews and you never read the original articles yourself, you will not have the knowledge base needed to counter-analyze and catch errors.

This creates a troubling paradox: Junior professionals or people new to a field, who might benefit most from AI assistance are also the least equipped to catch its errors and hallucinations. It is a common mistake in organization to allow people to use AI and experiment outside their purview, such as an accountant using an AI to start doing logos. **The AI works best as an amplifier for existing expertise rather than a replacement for developing i**t. You cannot effectively check something you do not understand, which means the AI becomes most dangerous not when it is used by experts who can verify its work, but when it's used as a shortcut to avoid developing expertise in the first place.

Finally, one of the best ways to use AI might be to challenge the exact opposite you wanted to do. Say your research indicates that you should target hospitals with your product rather than private clinics. Give that data to the AI and tell it "Based on this current data set, I am thinking of selling to hospitals. Show me I am wrong." Then read through the argumentations. It might identify a gap you have not seen. When challenging you, you could ask it, "Give me the Steelman counterargument," "Give me the stronger possible counterargument" or even "Give me a scathing critique." Of course, AI's counterarguments are not necessarily *correct* either; they are just forcing you to think through weaknesses and alternatives. The goal is not to accept whichever argument sounds more convincing, but to use the exercise to stress-test you are thinking and identify blind spots. Just be ready, AI does NOT hold back! But using the tool this way during analysis becomes a great way to identify previous analysis by either you or the AI that could be wrong.

As a practical example of this adversarial approach, I once gave both ChatGPT and NotebookLM the same dataset (patents, articles, and market reports) and asked them to review an inventor's contention that movement was not a challenge with their technology. NotebookLM agreed with the inventor, while ChatGPT disagreed. I had them essentially debate each other by pasting one AI's response into the other, and after about an hour of back-and-forth, NotebookLM conceded to ChatGPT's argument. The key difference was likely that ChatGPT could pull from knowledge beyond my curated dataset, being able to draw on outside data rather than relying exclusively on training data while NotebookLM was strictly limited to the documents I provided. This revealed a genuine gap in my original research: The dataset I would assembled did not adequately address movement concerns, and ChatGPT's ability to bring external knowledge identified a blind spot that could have been critical. Using multiple AI systems with different constraints to challenge each other turned out to be an effective way to stress-test both my data collection and the inventor's claims.

Finally, remember that many companies, particularly in healthcare, are justifiably cautious about adoption of AI due to the risk around data exfiltration and the lack of control over where and how their data is managed.[8] Just remember that in the end, you are the decision maker.

4.6 Obstacles to Effective Analysis

There are several issues a researcher must keep in mind when doing analysis. Doing so will play a key role in conducting data analysis in a coherent manner.

4.6.1 Confusing Facts and Opinions

When collecting data, the researcher will be faced with several different data sets originating from multiple sources. It is important to remember that not all of them will be facts and should be treated accordingly. A fact should be irrefutable. This can become confusing as you do online research and gather multiple statements that are in conflict: Multiple different market research firms can estimate market growth differently. It will be up to the researcher to select which one is the most credible and define it as such: Use words such as "estimate" and "believe" that convey confidence, while not falsely conveying confirmation.

A fact is something that has occurred, something that is true. Do not confuse opinions and beliefs with facts.

4.6.2 Avoid Falling in Love with Your Ideas: Managing Researcher Bias

One of my friends, a teacher, once shared that one of the hardest things about her job is not the teaching, it is the parent-teacher conferences where she has to say:

> *"There might be something going on with your child." "We start gentle,"* she told me. *Soft language, focus on potential. But by the first term, if there's really an issue, we have to pull the big guns. I'm often the first person to tell them. And its real hard. As parents, we all want our kids to be perfect. And it's never easy to hear. But it needs to be said.*

Market research works the same way. The uncomfortable truth might be that your innovation is not as perfect as you believe. That not everybody loves it as much as you do. Sometimes, as market researchers, we must be the ones to say: "There might be something going on with your technology."

This risk of tunnel vision can become especially dangerous when using AI for analysis, because AI responds to how you frame your questions. It is not objective. You might unconsciously craft queries that reinforce existing beliefs rather than challenge them. If you are convinced that doctors are the right target for your antimicrobial NanoPaint product, you might ask "Based on the interview data set supplied, explain how doctors are ideal decision-makers for purchasing hospital antimicrobial solutions" rather than "Based on the interview data set supplied, are doctors implicated in purchasing decisions for hospital infrastructure?" The first drives you down potentially the wrong road; the second gets you closer to the truth. The AI will dutifully build a compelling case for whichever premise you provide, making biased thinking look authoritative with proper citations and confident language.

Falling out of love with your ideas is difficult, especially after investing time, money, or reputation. One practical counter is to explicitly ask AI to challenge your assumptions. For example, request the "strongest possible case against my current thinking" or ask it, "What evidence contradicts my assumption?" This forces both you and the AI out of confirmation mode and into genuine analytical thinking. Finally, get other perspectives around decisions you are struggling with. As my friend added, "*When it does not work, we bring in specialists and professionals to help us on that journey. It is a long process, but when you care about the kids, you do what you need to help them.*" When you care enough about your innovation to see it succeed, do not hesitate to do the same.

4.6.3 Complexity of Data

As you engage in market research, complexity can become a challenge, especially if your data sources are increasing, and mixed in nature. You might find yourself in a situation where you need to reduce complexity. One way to do so is to try to simplify complex information and distill it into a single sentence. While you might lose some of the richness of the data, you can always go back to your original data (especially if your coding framework allows quick referencing).

4.7 Final Note of Data Analysis

This chapter dealt with data analysis, but with a very limited scope. We looked at the data that you collected and analyzed it, but it will be important to go beyond that data set to include macroeconomic data, tends studies and microeconomic factors, to name a few. It will be important to give a sense to your data beyond the limited scope of what you collected.

Notes

1 Huebner, M., et al. 2015. A systematic approach to initial data analysis is good research practice. *The Journal of Thoracic and Cardiovascular Surgery*, 151(1), 25–27. http://dx.doi.org/10.1016/j.jtcvs.2015.09.085 (Accessed January 22, 2026).

2 van den Broeck, J., Cunningham, S. A., Eeckels, R., & Herbst, K. 2005. Data cleaning: Detecting, diagnosing, and editing data abnormalities. *PLoS Medicine,* 2(10), e267. http://dx.doi.org/10.1371/journal.pmed.0020267 (Accessed January 26, 2026).

3 Manikandan, S. 2010. Preparing to analyse data. *Journal of Pharmacology & Pharmacotherapeutics*, 1(1), 64–65. https://doi.org/10.4103/0976-500X.64540. https://www.ncbi.nlm.nih.gov/pmc/articles/PMC3142762 (Accessed January 26, 2026).

4 Vach, W. 2013. Transformation of covariates. In *Regression Models as a Tool in Medical Research.* Taylor & Francis Group: Boca Raton, FL, USA; 264–273.

5 Bivariate analysis refers to the analysis of two variables. If three or more variables are analyzed, then it is called multivariate analysis.

6 Williams, K., & Page, R. 2011, April. Marketing to the generations. *Journal of Behavioral Studies in Business*, 3. http://www.aabri.com/manuscripts/10575.pdf (Accessed January 28, 2026).

7 To illustrate this point, I decided to test the very example I'd just written. I asked Claude, "What is the best place to eat dholl puri in Montreal?" It returned several recommendations, including La Cuisine de Tatie in Brossard, which I happen to know is one of the best restaurants in the region for Mauritian food. So far, impressive! The AI had correctly identified a genuinely authentic Mauritian restaurant from a niche query. But when I looked closer at the recommendation, something was off. Claude had attributed it to an academic article titled "Human–machine collaboration in transcription" from the *Journal of AI, Robotics & Workplace Automation.* The article had nothing to do with Mauritian food, it was about speech recognition technology.

What happened? The AI gave me an accurate recommendation but hallucinated the source entirely. It might have found the restaurant through legitimate reviews, but when assembling its response with citations, it grabbed a completely unrelated source from the search results and confidently presented it as the basis for the recommendation. This perfectly demonstrates a core problem: AI can produce useful outputs while simultaneously fabricating the evidence trail, and unless you have independent expertise to verify both the conclusion and its sources, you won't catch the deception.

8 Dupin, M., & Oruç, M. F. 2025. Enhancing market research with a GPT-based API wrapper: A leap towards advanced data analysis. *Applied Marketing Analytics,* 10(4), 348–373. https://doi.org/10.69554/AWCT1053 (Last accessed January 28, 2026).

Chapter 5

Models and Frameworks

The more you learn, the more you have a framework that the knowledge fits into.

—Bill Gates

Data must be turned into a story to be compelling. Hence, you must work on highlighting an actionable insight or distilling it into a model that will help the recipient to understand my message. After collecting and analyzing the data, I work to integrate the data into a framework that eases comprehension and emphasizes the most important elements. In the following sections, we will be going over the SWOT model (and its alternative, the SCORE model), the TAM-SAM-SOM model, the Kano model, and the PMF (the product market-fit model).

5.1 Appraising the Market Environment: The SWOT Model

The SWOT model is a simple yet clear framework used to present your organization (or your product or service) in relation to its environment. It is also possible to build SWOTs of competitors and their products.

In a SWOT model, the strengths and weaknesses refer to the organization itself (internal elements), while opportunities and threats focus on the environment outside the organization (external elements). Hence, building a SWOT model allows us to define the organization comparatively to its competitors, identify the best opportunities for growth, and identify the threats (current and potential) inhibiting potential growth.

DOI: 10.4324/ 9781003731726-5

The main advantage of the SWOT model is that it is easily recognizable: Most individuals have seen a SWOT model at one time or another and your recipients will be able to recognize it with a quick glance, letting them capture the information you are conveying. Its simple nature lets the user condense and simplify complex data. This is also its disadvantage, as a tendency to oversimplify complex situations can leave information off the table. Another challenge is that every factor is weighted equally in a SWOT model, so it is impossible at first glance to determine which factors are crucial, and which are ancillary. Finally, SWOTs are subjective, so expect the building of your SWOT model to be iterative, making changes as more information becomes available and more feedback is obtained.

Before engaging in a SWOT analysis, it is very important to clearly define the objective of the SWOT model. As we will see in the following sections, many components are subjective, but having a clear opening statement and a vision of what you are trying to achieve will help you correctly categorize your information.

5.1.1 Four Elements of a SWOT Model

The four elements of a SWOT model are divided into two broad categories: Internal elements (strengths and weaknesses) and external elements (opportunities and threats). In general, internal elements are those that you have some measure of control over, while external elements are those which you have no control over.

Internal attributes are those that are unique to your organization. They are the elements that define you and differentiate you from your competitors. You learn this by analyzing your internal resources, such as employees, board of directors, and partners, as well as financial and sales data. You could also include a scan of your organization online, identifying how your product or service is rated or perceived, as well as a comparative benchmarking of leading products.

External attributes include all third parties that are outside your organization that impact your organization. Regulations, laws, suppliers, and competitors are all part of the external environment that directly impacts your organization (Figure 5.1).

Strengths are positive elements that give your organization a competitive advantage or unique value over competitors. A specific technology, an

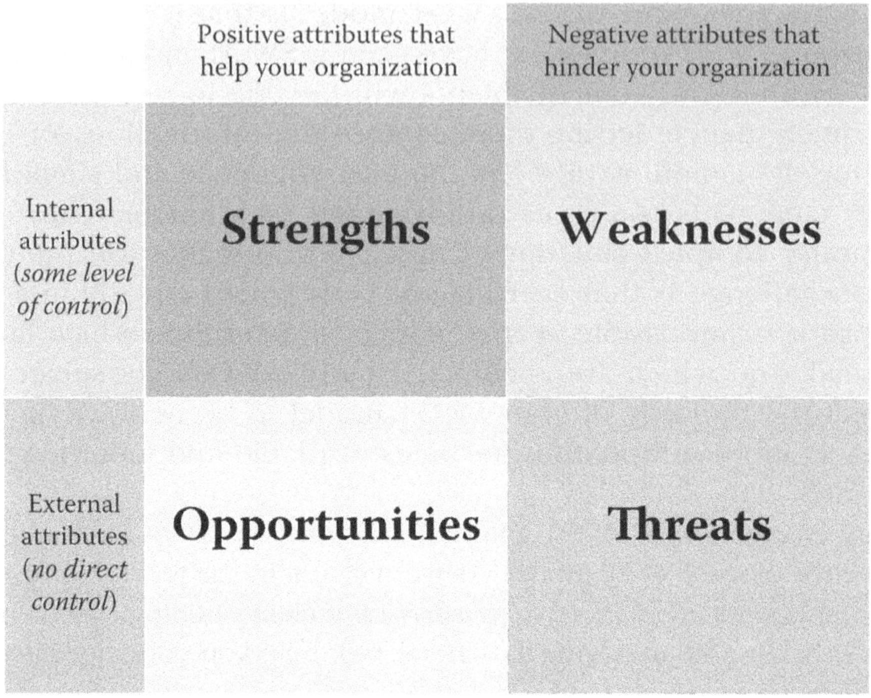

Figure 5.1 SWOT model framework.

exclusive license, and a strong patent portfolio are all unique attributes that can serve to distinguish you from existing competitors. Team members with unique skills and experiences are also strengths that need to be showcased. Some of the questions that can guide you in building the Strength element of your SWOT model include: What can we do well? What is our unique expertise? What unique resources do we have access to? What advantages do we have over our competitors?

Weaknesses are the attributes that are inhibiting growth in your organization. They are the attributes that place you at a disadvantage compared to your competitors. For example, using a legacy production system to build your product, a lack of sales expertise, or not having a patent to protect your innovation are all weaknesses that can inhibit your growth. Some questions you can ask yourself when defining your weaknesses include: What is it that we are not good at doing? What are our competitors doing better than us? Is there anything we do that we know we could improve?

Weaknesses can be difficult to define. Therefore, it may be necessary to work with external stakeholders (e.g., clients, suppliers and consultants) to help you define them. Many organizations are wary of facing their own internal weaknesses or believe that drawing attention to these weaknesses will be a potential strike against them during investor pitches. The important thing to remember is that defining weaknesses is the first step, but its important information to highlight is how the company will address them. As such, recognizing and addressing a weakness is a much better alternative to hiding it under the rug, hoping nobody will notice it.

Opportunities occur when there are changes in your environment that you are uniquely positioned to take advantage of. New regulations, exclusive access to new suppliers, or the difficulties of an important competitor all generate opportunities for your organization. Opportunities can emerge in technologies, social and lifestyle trends, population changes, government policy, markets, and more. As such, continuous monitoring of your environment is invaluable to identifying opportunities. You might also decide to engage in activities that directly lead to opportunities such as identifying grants, developing partnerships, participating in trade missions or attending conferences and events.

Some questions that can guide you in identifying opportunities include: What are some of the trends impacting my client base? What are some of the major new regulations that have impacted my industry in the last few years?

Threats are shifts outside your organization that can have an impact on your organization. While they are outside your organization (meaning, you have little control over them), recognizing and assessing them can help you prepare your organization, and identify solutions to address them. For example, a shift in regulatory procedures could incentivize you to change some elements of your production model and look for appropriate subsidies and partners to help you make the necessary adjustments.

You might not be able to change the external environment, but you can react once the change occurs (or even better, you can plan to adjust before the change occurs). Some examples of threats include new competitors entering your market, current clients shifting their strategy, new technologies affecting consumers' purchase and use of your products, changes in social trends, government regulations, or changes in the availability of crucial supplies. Some of the questions you can ask yourself are: What are my competitors doing? What new technologies are being developed? What are some of the regulatory components that are under review or discussion?

5.1.2 Developing Strategy Applications from Your SWOT Model

There are two simple approaches to developing a strategy using the information in your SWOT model: Matching and converting.

Matching implies the pairing of your positive attributes (strengths and opportunities) to identify prospects for your organization. For example, if one of your strengths is a unique production process, and an opportunity emerges in lower regulatory hurdles for a specific market, you might then identify that market as a potential new market to target with your product as you are able to scale up rapidly.

Converting means identifying a weakness or threat and transforming it into a strength or opportunity. For example, if a competing startup is developing an innovative technology that threatens your market, this might be an interesting opportunity to partner up and co-develop said technology. Other opportunities could be developing a similar innovation or partnering with a competing innovative technology to develop a new *unique selling proposition.*

5.1.3 An Alternative to SWOT: The SCORE Model

An alternative to the SWOT model is the SCORE model. Like SWOT, it is a strategy framework used to develop a quick assessment of your current situation. SCORE stands for Strengths, Challenges, Options, Responses, and Effectiveness. SCORE is less of an environmental scan tool, but rather a problem–solution model. You identify your strengths and the challenges you are facing and start developing solutions to address them. As such, it is more of a road map toward problem solving than SWOT is and includes a measurable component to ensure you are on track.

The key to using SCORE is to select a challenge and start developing the framework. Assess each item as you progress and identify any measurable items.

Finally, remember that while SWOT is somewhat static, SCORE is an iterative model, and can be re-examined on a recurring basis. It is a useful model to demonstrate the acknowledgment of issues, as well as a road map to resolution. To assist you in building your own SCORE model, you can use the SCORE framework development table included (Table 5.1).

Table 5.1 SCORE Framework Development

SCORE Attribute	*Question Guide and Objective*
Strengths	• What are my current capabilities? • What are my current internal resources? • What are some of the outside resources that I can call on? *Objective: Inventory of capabilities*
Challenges	• What are some of the issues I need to address? • What are the capabilities/resources I need to acquire/ develop? *Objective: Inventory of issues*
Options	• What are the options to address these challenges? • What is the risk associated with each option? • How should we prioritize them? *Objective: Strategy scenarios*
Responses	• What are the expected responses to our options? • From regulators? Competitors? Stakeholders? • What return/reward do we expect from each option? *Objective: Risk-management scenarios*
Effectiveness	• For each option • Is it effective? • Is it realistic? • Is it reliable? *Objective: Project management assessment*

5.2 Appraising Market Size: The TAM-SAM-SOM Model

To appraise your market size, a TAM-SAM-SOM model is an interesting approach since it combines elements relating to both the top-down market sizing approach (sizing the market size from the literature) and bottom-up approach (estimating the market size from the perspective of the organization's available resources).

There are multiple advantages to this model. First, it is relatively easy to use and explain. Second, it combines the top-down and bottom-up parts of market research, which are essential to telling your story. Finally, a well-built TAM-SAM-SOM demonstrates a solid understanding of the market and it is limits.

The main challenge with TAM-SAM-SOM model is that it relies on the strength of your assumptions. As such, building a solid model will depend on a realistic assessment of the market and of the capabilities of your organization. Furthermore, the justification behind your commercialization plan will be crucial in justifying your SOM.

5.2.1 Three Parts of the TAM-SAM-SOM Model

The TAM-SAM-SOM model is divided into three distinct parts that are imbricated into one another. The largest part is the TAM, which is the *total attainable market* for a specific product or service. It contains the SAM, which is the *serviceable available market,* the market that is within reach of your company. This in turn contains the SOM, which is the *serviceable obtainable market,* the realistic market share that you are likely to obtain, or the one that you plan to achieve in the short term (Figure 5.2).

To properly illustrate, let us use the very simple example of a small biotechnology company developing a diagnostic device in the small cell lung cancer segment.

- The *total attainable market* is the market from its broadest perspective. For our small startup, the TAM is the total lung cancer diagnostic

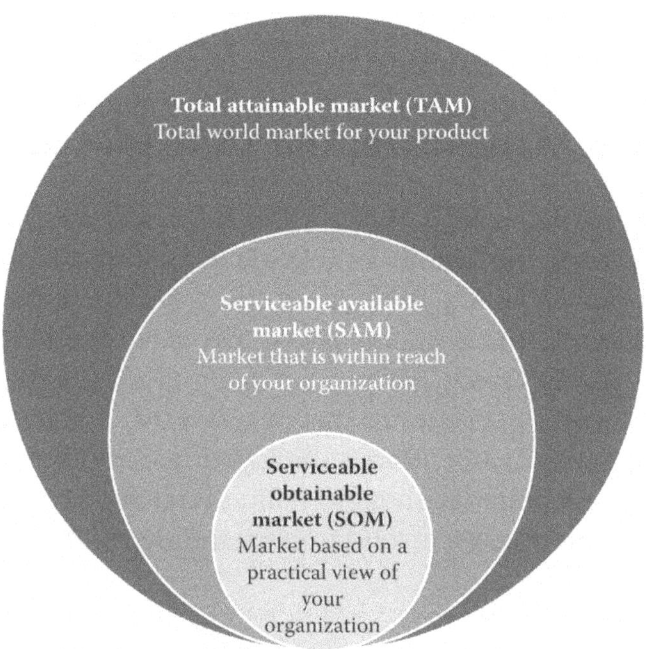

Figure 5.2 TAM-SAM-SOM model framework.

market space. In this regard, Grand View Research estimated that the global lung cancer diagnostics market revenue is expected to reach $3.6 billion by 2024.[1] This could be seen as an approximation of the future total available market worldwide. But a company can seldom claim this to be their target market. The resources necessary to target this total market simultaneously are quite ambitious. So, the next step is to calculate the serviceable available market.

- The *serviceable available market* is the area that is within reach of your company. For example, in the case of our small startup, we can make a series of assumptions:
  a. The diagnostic device specifically detects small cell lung cancer, which accounts for around 15% of the lung cancer diagnostic market share.
  b. In the short term, your company will be targeting mainly the North American market, which accounts for 30% of the revenue share. Hence, your serviceable available market share is estimated at $162 million (Table 5.2).
- Finally, you start looking at your *serviceable obtainable market,* which is the realistic market share that you can likely acquire, or the one that you plan on achieving in the short term. This data has to be backed by a bottom-up forecast, detailing how you will utilize resources to achieve your revenue goals. Without going into the details of a commercialization strategy, let us imagine that you have made a 2-year plan that generates $10 million in sales.

Once you have developed your model and data, it becomes quite easy to demonstrate your market with a bubble graph as shown in Figure 5.3.

As you can see from Figure 5.3, this type of modelling depends on the series of assumptions. Your market research will be the cornerstone of your results. A good understanding of the market and the competitors is essential to building a justifiable SOM, and recipients will be as interested in the

Table 5.2 Sample Serviceable Available Market Development

Estimated TAM	$3.6 billion
Estimated market share of small cell lung cancer segment	15%
Estimated market share of North American market	30%
Estimated SAM	$162 million

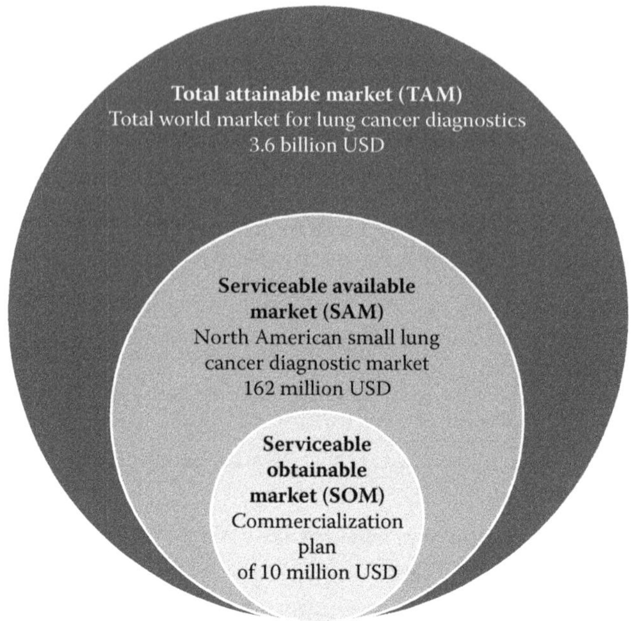

Figure 5.3 TAM-SAM-SOM model for a small cell lung cancer diagnostic device.

numbers as how you developed the model: Was the presenter realistic when he built his model, was he pragmatic? Remember that, in general, forecasting a SOM that is greater than 10% of the SAM is going to require a great deal of justification.

Also, as with everything else, TAM-SAM-SOM models are iterative, and can be adjusted as new data becomes available. Finally, it is always a good bet to be a bit more conservative with your numbers: It is easier to explain how you beat your forecasts than justifying why you fell short of the number that you promised.

5.3 Appraising Customer Attributes: The Kano Model

Your customers' appreciation of your product attributes can be classified following many different models. In this section, we will be looking at the Kano model of attribute classification. The Kano model classifies purchasing attributes into five distinct categories: Threshold attributes, performance attributes, excitement attributes, neutral attributes, and reverse attributes.

The Kano model is a useful framework to classify customer perceptions and enables readers to quickly identify which features will generate the higher returns (since they are linked to creating higher satisfaction), and which features are not worth investing in (since they are not creating any satisfaction, or even worse, they are generating dissatisfaction). Take note that these attributes do migrate over time. The longer an excitement attribute has been promoted, the more it becomes a performance or threshold attribute. We will be going over each attribute, followed by a short discussion on the limits of the Kano model, which includes the dimension of time.

5.3.1 Five Attributes of the Kano Model

As mentioned in Section 5.3 there are five distinct purchasing attribute categories in the Kano model: Threshold attributes, performance attributes, excitement attributes, neutral attributes, and reverse attributes.

Threshold attributes are the attributes that the consumer expects in a product. They must be present, and since the consumer anticipates them, they do not usually create a differentiation opportunity. Increasing the focus on or investment in these attributes does not usually provide more sales or greater client attraction, but removing or diminishing these attributes will result in extreme customer dissatisfaction. In other words, the customer expects that when purchasing the product, their basic needs will be fulfilled, and that this attribute will be present to fulfil that need. A simple example of a threshold attribute is the absorbent pad on an adhesive bandage. The client expects that there will be an absorbent pad on the bandage; the lack of a good absorbent pad on an adhesive bandage will turn clients away, and it would be difficult to focus on this pad as a unique selling point.

Performance attributes are those where the more the attribute is present, the better it is for the consumer, and it will improve customer satisfaction. They are clearly expected by clients. The price a customer is willing to pay for a product is closely tied to performance attributes. Nonetheless, products that focus on performance attributes are interchangeable with other products focusing on performance attributes, so it is unadvisable to build a product strategy that focuses exclusively on performance attributes. A performance attribute in an adhesive bandage could be a material that makes the consumer more comfortable or a material/glue that makes the adhesive stay on the finger longer.

Excitement attributes are those that are unanticipated by customers, but which create high levels of customer satisfaction. However, their absence does not generate dissatisfaction. These are the attributes that enable you to differentiate your product from products with similar performance attributes. Providing excitement attributes that address "unknown needs" generates a competitive advantage over competitors. Following our adhesive bandage example, bandages with antiseptics directly in the absorbent pad or bandages that help to regenerate wounds could be examples of excitement attributes for consumers.

Neutral attributes are attributes for which the customer does not manifest a distinct opinion. Their presence neither improves nor impedes the purchasing decision. Frequently, these are features that excite the innovator but are seldom of interest to customers. Hence, it does not mean that the attribute does not add value to the product, it just does not add any value for the consumer. For example, the color of the adhesive generally does not influence a purchasing decision (unless you have chosen an especially hideous color, in which case you might have unknowingly generated a reverse attribute), but it might be easier for a manufacturer to privilege and produce their product in a single color.

Reverse attributes are rather rare and occur when a client perceives an attribute negatively. In that case, he will be willing to pay an amount of money to have them removed. As such, the presence of a reverse attribute creates a negative experience, while its absence creates a positive experience. For example, some clients might be willing to pay more for plain adhesive bandages, rather than buying cheaper adhesive bandages with kids' cartoons on them.

The Kano model is a simple but powerful model, useful in identifying opportunities in customer needs, as well as demonstrating how these attributes impact the consumer's purchasing decision. The more the attributes create customer satisfaction, the more effort should be invested in developing and refining these attributes.

5.3.2 Issues with Kano Modeling

There are two issues with Kano modelling. One is the impact of time on attributes; the other challenge relates to different consumer perceptions.

The main challenge with Kano modelling is the impact of time on the perception of consumers. For example, an attribute that was initially classified as an excitement attribute can quickly become a threshold attribute as more companies integrate the target feature. For example, think back to a few years ago when cameras and MP3 players were integrated into mobile phone devices. While this was once an excitement attribute (*Wow! I do not have to carry three devices anymore, everything fits onto a single device!*), could you now imagine purchasing a mobile device that could not play music, or could not take pictures? As such, the Kano model needs constant reappraising to ensure that attributes are correctly classified, and that the market has not shifted away from some of your previous assumptions.

The second challenge is that you can find inconsistencies among consumer perceptions. As you start collecting data, you might realize that while some attributes are a threshold attribute for some individuals (a camera is a must for some potential clients), it becomes a negative attribute for others (*I have to pay extra for a camera? But I never use it! Show me a cheaper model without a camera instead!*). Careful segmentation of your data and client targeting alleviates this challenge: As you precisely identify your customer segment, you will eventually distinguish patterns between them. The clearer your end market, the less likely you will have inconsistencies in consumer perceptions.

Finally, for most customers, they are researching on their own, talking to peers. It is only late in the process that they will start interacting with. By that time, they will have well-informed opinions and will have decided on what is important to them. As such, your ability to influence how they perceived attributes will already be set.

5.4 Product Market Fit

For an entrepreneur, demonstrating that their innovation is right for the market and will generate interest is a vital step to raising funds. As such, it is important to have a framework that lets you validate if your product is a good fit for the market. One of the frameworks that has rapidly gaining popularity is the Product-Market-Fit (PMF).

PMF is used to validate that your product's value proposition will generate actual customer interest to purchase your product, which will translate in real sales. This step usually follows the Problem-Solution-Fit (PSF), which is used to identify the best solution to solve a problem. The difference between the two is quite noticeable: If you did a PSF, you might have developed a product that is the perfect solution to a problem, but you might not find anybody willing to buy it (due to cost, complexity, and so on). As such, measuring PMF ensures that you have demonstrated the **market intent** in purchasing your product.

5.4.1 Evaluating Product Market-Fit

So how do you know you have a good market fit? One interesting framework to evaluate your product market-fit was developed by Adam Fisher, of Bessemer Venture Partners. He uses it to quickly evaluate companies in his portfolio as well as for founders to reflect on their own product-market fit journey.[2]

On the x-axis, we measure the depth of customer engagement, which we uncover by actively talking to their potential customers, understand their needs and current concerns. Entrepreneurs that are either out of touch, or talking to a limited number of stakeholders who are already convinced of their technology might not be challenged and as such might believe they have a good grasp of their customer but are stuck in what he calls an echo chamber.

It can be challenging to balance between integrating customer feedback and sticking with your vision. Obviously, you do not want to shift your product to every feedback, but then again, it can be difficult to decide which feedback to listen to. Nonetheless, there are some things you can do. First, take care to notice patterns in feedback. A lone potential customer might not be a great indication but if you are able to guess what the customer is about to say before they end their sentence, it might be time to seriously consider the feedback. I had a startup I was working with, and I was sharing some feedback on their product, to which the founders responded "Yeah, everybody says that, because they do not understand what we are trying to do." In this case, it might be the reserve that is occurring, where the founders are not listening to the feedback, and refusing to adjust their product.

Other things you could do is dig deeper with the persons sharing feedback (You mentioned feature X is essential; can you tell me why? How would you use it? Do competitors already do it? How does it help you accomplish you work?) or contacting an independent third-party expert to discuss (multiple companies have mentioned that X is especially important, what is your experience in this space?) For example, I was working with a university professor that had developed an interesting software for diagnostics. Discussions with doctors resulted in a lot of pushbacks, as many did not think this technology would integrate well into existing ERPS. To get another side of the feedback, we contacted Hospital IT administrators to get their perspective from a technology perspective, and the results were quite valuable in the commercialization strategy.

On the y-axis, startups measure how compelling their product is for potential customers. Is it generating buzz? Is it challenging conventional thinking? Is it garnering strong interest? A startup on the lower end of this axis might not have an inferior product, just one that does not really resonate with their customers.

It might be difficult for a company to measure the resonance of their product. Here, you can use the 40% rule: If at least 40% of the potential customers you talked to expressed strong interest in your product as something they absolutely need, then it is a good indicator you have good resonance. If you are lower, you might struggle to gain significant traction to reach your market. You might need to redefine your target market, or you might have to redefine your product.

The obvious question that comes next is "How do you measure strong interest?" Once again, nothing formal exists, but there are clues that you can gather. If you ask a potential client "So are you interested in my product?" and you get a positive response, it is difficult to evaluate between polite interest and real traction. Some better indications could include the client taking the lead in the interview with questions such as "This sounds really interesting, when do they expect this to come out?" or "This sounds like a cool product, are they looking for beta users? I would love to setup a trial at my organization!" In a product market-fit project I did, a product had such resonance that potential clients contacted me a year after the interview, inquiring on the technology progress and availability of trial products . . . that's strong resonance!

Another way you could measure resonance is asking if the interviewee agrees to be contacted in the future as the product nears the market. When I do this for client, I always wrap-up calls by asking "Would you like to be contacted by the company once the product nears the market to get a more detailed description of what the product does?" If I have a 50% positive response rate of interviewees agreeing to share their coordinates to the client, I know that we have good resonance.

If you are already selling your product, some practical hints that you have not achieved resonance are that word of mouth is not really spreading around your product, usage is not growing significantly, media and press interest around your product is weak, the sales cycle is too long (and it is growing longer), and many deals just never close. Inversely, if you are hiring sales and customer staff as fast as possible, sales are faster than your production, and reporters are calling to learn about your product, then you have achieved resonance.

Figure 5.4 The four quadrants of product market fit.

Coming back to our model, once you have measured both resonance and customer engagement, you can use the framework in Figure 5.4 to evaluate where you are in the product market-fit journey.

If you have high resonance with your market, and high customer engagement, congratulations, you have successfully achieved **product market-fit!** The next challenge will be maintaining yourself in this quadrant, watching out for challenges such as high-customer churn (which would suggest you are growing out of touch with your customers).

Many startups fall into the sexy story or revenue-centric quadrant. Having generated early sales, they figure they have found a good fit, but these customers might be outliers rather than being a manifestation of your main customer base. You might be getting the wrong feedback, or a very niche segment of the market might be excited by your product.

Companies that are **revenue-centric** are generating sales, but this is masking that they have not attained customer market fit. For example, each sale might be very customized with a high-level of post-sale service, or it could be having a very wide range of different clients. It might be fine in the short term as you bootstrap your company, but it will be difficult to scale up.

Companies that are **sexy startups** often do not realize they are in this position until it is too late. They might have an interesting product, that resonates well with media and investors but find themselves unable to commercialize beyond their early adopters or are unable to monetize their early free users. While they have great feedback, they do not realize they are in an echo chamber, and that the sales that do come through have a prohibitive cost of sale per unit or are irregular. These companies substitute customers with media and investor attention, assuming customers will eventually join in.

The last quadrant is one all too familiar to experienced investors and stakeholders in the startup space. We call these companies **self-deceiving startups.** These companies have neither customer resonance, nor no customer engagement. This is typical of small early startups where the company is being driven by founders who are convinced that their technology is essential, and that eventually customers will "get it." These companies usually refuse to adapt to customer feedback and normally do not get past the early funding rounds.

Another way of framing the discussion is looking at which type of team you have. If you have an engineering driven team, you risk putting yourself in the self-deception quadrant. If sales are driving your team, that you risk becoming a revenue-centric company, and if its marketing that is driving your company, you could become a sexy startup. The ideal situation is a product driven team, with assets in all other three quadrant. This team is usually best placed to grow a successful startup.

5.5 A Final Note on Organizing Complex Data

This section is dedicated to describing a few formal and well-known models to help organize your data. They do not apply to every situation, and there are many others we did not go over that could be of use, such as Porter's Five Forces Model for competitive strategy or the PEST (political, economic, social, and technological) analysis for understanding how outside environmental factors affect the organization. Two other books in this series, The Handbook of Marketing Strategy in Life Sciences and *Innovate, Fund, Thrive* go into more detail about how to use your data and build different models such as building a customer persona or completing a competitive environment analysis.

Nonetheless, when organizing data for presentation, there are a few things I try to keep in mind. First, when reviewing my data, I look to make sure that it is logically structured. Look for opportunities to showcase symmetries (these are usually well understood) and make sure to showcase clear connections and flow. Also, take a look at your data, and try to identify any unimportant data that you can remove.

Afterwards, take a step back to look at the big picture and the details to see if the point you were trying to make came across clearly. It can be useful to ask someone else to review your data, just to make sure that the point is consistent and clear, so ask for a second opinion. The collaboration that results enriches the presentation and produces better overall analysis.

During that time, focus on your message. It is quite easy to lose your main message when analyzing complex data, so make sure to compare your results to your initial message to ensure there is a match between them.

Notes

1 Grand View Research. 2016, June. *Lung Cancer Diagnostics Market Worth $3.64 Billion by 2024*. https://www.grandviewresearch.com/press-release/global-lung-cancer-diagnostics-market (Accessed January 28, 2017).

2 Fisher, Adam. 2019, November 25. *Adam Fisher on how to Navigate the Product-Market Fit Journey*. https://www.bvp.com/atlas/the-product-market-fit-journey#page_top (Last visited May 27, 2023).

Chapter 6

The Look 'n' Feel

People don't know what to do when you give them too much information.

—Caroline Florence

In 2025, I attended a presentation during BIO. Figure 6.1 is one of the slides that was shared. It was displayed for less than 30 seconds.

The level of information in this single slide is . . . immense. And as it was displayed for less than 30 seconds, it is somewhat unreasonable to expect the viewer to retain any amount of information.

I believe we all have had the opportunity to see slides like this that obfuscate information, rather than communicate it. Sometimes the slides are made of a "solid wall of text" (a dozen or more full sentences written in a 16-size font), which the presenter promptly reads like he was reading a chapter from a book. Sometimes the color contrasts make the text illegible, spelling mistakes make the text unreadable, or badly labelled graphs are displayed, which are impossible to decipher.

These presentations failed, not because of the information in them, but rather because the people presenting them failed to turn data into consumable information.

When presenting, the look n' feel of your data is a crucial element to ensure that the audience captures your message. Mishandling the presentation elements can drown your message, and the intended recipient will be unable to receive the information you are attempting to share the way that you meant them to. Even worse, the recipient could interpret it

DOI: 10.4324/ 9781003731726-6

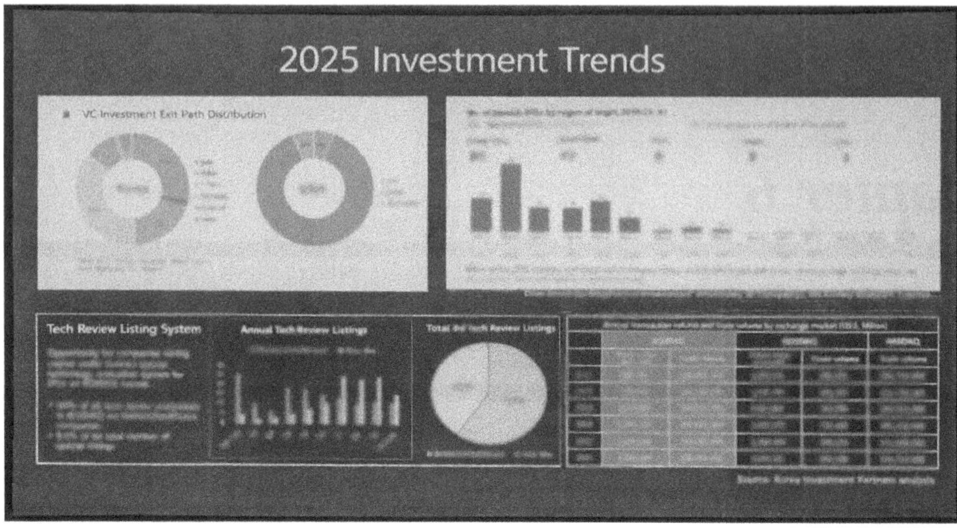

Figure 6.1 Example of a slightly overloaded PowerPoint slide.

completely differently to what you were trying to convey it. Presenting data with attractive and clear visuals is essential to sharing the results of your data collection and analysis, as well as helping you make better-informed decisions.

Transforming data into eye-catching graphics and tables is not as intuitive as it may seem. While new software and platforms make the mechanical aspects of data presentation simple to master, everything from choosing the right graphics to the graphical design aspects are a unique set of skills, skills I refer to as look n' feel skills. They include choices related to the look of the presentation (colors, fonts, shapes, layout, and graphics) and the feel of the presentation (architecture of the graphics and tables, responsiveness, and interactivity). Look n' feel also includes the effort to create a consistent image across a presentation, and the overall branding image that is generated.

This chapter will be dedicated exclusively to communicating our market research findings to our audience. The key to a good graphical presentation is to select the method that best fits the data, so we will start by going through a process to choose the best way to communicate quantitative information. This will be followed by information on models for presenting qualitative data. The chapter will conclude with an overview of presentation tools that integrate quantitative and qualitative data, such as slideshow tools, the use of infographics and a few tips of storytelling itself.

6.1 Presenting Quantitative Data

The transformation of quantitative data into a comprehensible output is a necessary step to getting your message across. To do so, it is important to decide which graphical layout is the best one to demonstrate your information.

- *Tables are used when direct access to precise numbers is necessary to support your message:* In a *table,* information is displayed across rows and columns. A table is useful if precise individual values must be shared. For example, imagine doing a presentation to illustrate demographic data relative to your product: If sharing the precise market share percentage for each demographic segment is crucial to the presentation, the use of a table is indicated. Ideally, a table should have a clear title, include information such as the sample size and when the data were collected, and is inserted after the text that relates to it.
- *Graphs are indicated when a relationship in data is what you want to demonstrate:* A *graph* is data that is displayed in a visual layout, such as a pie chart, a line graph or a histogram. Visualized data allows you to demonstrate information in a relationship to another data set, along one or more axes, and is useful when you are trying to show the shape of data, its patterns and its trends. Following the previous example, if the message you wanted to circulate is how much your product is sold to each demographic segment in relation to other competitors, then the use of a chart (e.g., a pie chart) effectively communicates relative market size from one company to another. Ideally, a good graph will include a title, a clear scale and clear axes so the reader can quickly understand the data.

Note that if you are showcasing a single important point of information or one data point, simply use large tile or label for your data, rather than embedding it in a table, a graphic or a visual component.

6.1.1 Transforming Quantitative Data into a Graphic

If you have determined that a visual layout is needed for your information, you can identify which graphical representation is the most pertinent to communicate your message. Remember that "fancier" does not mean better.

Sometimes, a simple line chart is all you need to properly convey the information you are trying to share.

A simple approach to selecting the best visual layout for your data is to identify the data relationship you want to demonstrate. Once you have decided the type of relationship you want to profile, you can use Table 6.1 to narrow the best graphic representation for your data.

Table 6.1 Choosing the Most Relevant Graphic Representation for Your Data According to the Type of Relationship

Relationship	*Description*	*Graphic Representation Alternatives*
Time	You are observing a single variable over time	• Use a *Line Chart* to emphasize the shape of data • Use a *Bar chart* to emphasize contrast between data sets
Ranking/Score	You are scaling items in order during a single period	• Use a *Bar Chart,* especially horizontal bars if you have a long label to place • Use a *Radar Chart* if you are comparing scores across multiple objects
Part of a whole	The items you have analyzed are part of a whole	• Use a *Pie Chart* when comparing one key item versus the whole market • Use a *Bar Chart* when comparing trends across attributes is important • Use *Stacked bars* when comparing whole categories across time
Benchmarking	Items are compared to a reference item	• Use a *Bar Chart* to compare your data to benchmarks • Use a *Line Chart* to emphasize shape of data
Frequency	You have compiled several observations per interval	• Use a *Histogram* to emphasize individual values
Correlation	You are comparing two (or more) variables across time	• Use a *Scatter Plot* to show how one attribute is affected by a second variable • Use a *Bubble* to show data from a scatter plot, emphasizing a third variable

As a broad example, many years ago I wrote an article for a marketing magazine.[1] Using data from a client's quinquennial survey, I wanted to explore the relationship between a product's environmental attributes and consumer purchasing patterns, as well as measuring how these patterns evolved over time. The hypothesis was that over time, environmental attributes would become more important in a consumer's decision process.

However, the data articulated a completely different story, as there was no significant difference in consumer purchasing patterns between 2009 and 2013. The importance of environmentally friendly attributes and the willingness to pay more remained at the same levels between the two surveys. As I was demonstrating multiple relationships in data (multiple time periods versus purchasing interest), I used a simple bar graph (Figure 6.2) to illustrate the data. This way, the reader could clearly see that the consumer levels remained constant from one data set to the next. The use of a graph was indicated because (1) I was illustrating a trend and (2) the relationships between time and attributes were key to understanding the story.

6.1.2 Building a Graph

Once you have identified which type of visual layout you wish to build, and have started design, you can use the roadmap below to be sure you are not forgetting any information.

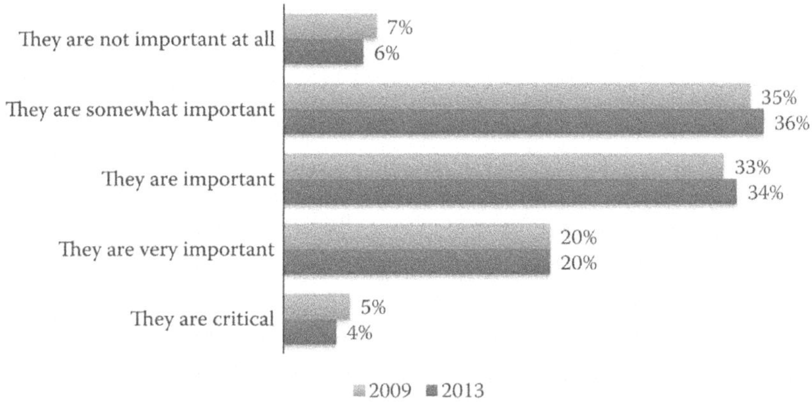

Figure 6.2 Using a bar graph to illustrate "How much influence environmentally friendly attributes have on your purchasing decision."

A. *Before building the graph*
   - If you are doing a graph with mixing quantitative and qualitative scales, place the quantitative scale on the Y axis (vertical) and the categorical scale on the X Axis (horizontal); time is especially important to place horizontally since it follows our intuitive habit of seeing time moving from left to right.
   - If you use a bar chart, your horizontal scale will most usually start at zero, but if you are using a line chart, you can narrow the scale and starting the data at a more relevant number. This will give the reader a better sense of the pattern or story you are trying to convey.

B. *Once you have built your graph*
   - Remove any distractions that have crept into your story. When building your graph, focus on the data itself, and try to remove any graphical elements that aren't helping to understand the message you are conveying. For example, the use of grid line is not recommended unless they are useful for the reader to interpret data (ex. the reader needs to precisely see where the data is)
     - *A note on color coding: Use softer natural colors when building graphical elements, while saving more flashy colors for items you need to stand out. Once you determine a color scheme, stick to it for the entire presentation, and* ***do not shift from one color scheme to another as the presentation advances, as it becomes an additional distraction.*** *More information on colors is available in the section on slideshows (6.3.1).*
   - Do not forget to include all relevant descriptive text: The graph title and the axis titles are invaluable for the reader to understand the message you are communicating, as well as the legend that further contextualizes the data.
   - Decide if you need to emphasize a specific data set. If one element of data is more important than the others, highlight it accordingly, either using bold colors, or using a different graphical item.

6.1.3 Decision Tree Modeling

Decision tree modelling is a visual presentation model used to illustrate a decision process and their corresponding actions. It can be used to illustrate any process with choices such a purchasing process or the impact of a marketing campaign. The top of the tree refers to the choice alternatives, followed by the decision criteria and closing with the decision outcome.

As an illustrative model, decision trees have many advantages. First, they are easy to understand, as it they are a visual representation of a phenomenon, and it is very intuitive to follow. They are also very useful to enhance decision-making, or as a starting point to brainstorm. For example, it could be used to illustrate a consumer decision process. It also allows to illustrate how one data set impacts a second one.

In another example, a client in the dental supplies market was looking to offer a new service and was interested in determining how much did delivery model impacted dentist interest in purchasing in relation to the price she was targeting. As such, a decision tree (Figure 6.3) was modeled to help illustrate the collected data from a survey.

Using the data collected in the survey, this decision tree model was built. The choice was around the purchasing decision at a certain price point, the criterion was the delivery speed, and the outcome was the purchasing decision. As we can see, the slower the proposed delivery model, the lower the purchasing intent.

Overall, decision trees can provide insight into decision process, enable better decision making and illustrate to the audience the data behind a specific decision.

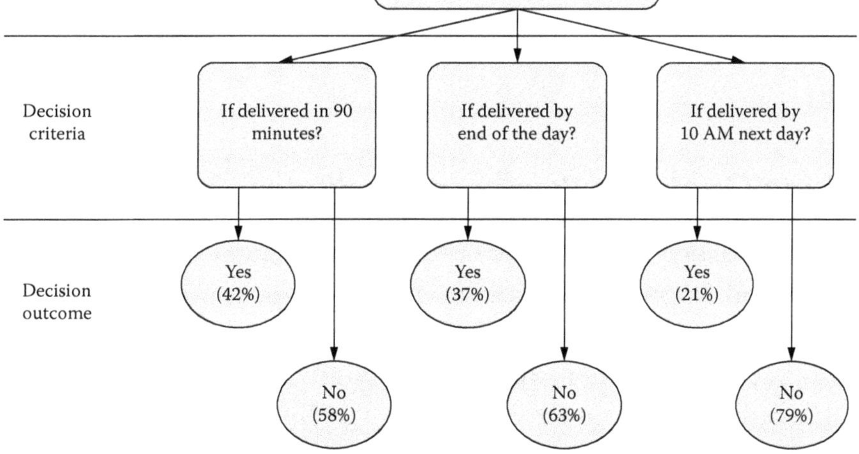

Figure 6.3 Decision tree modelling example showcasing how price, delivery model and purchase intent are interrelated.

6.2 Presenting Qualitative Data

Qualitative data differs from quantitative data during analysis, a difference that persists all the way through the presentation of data. The richness of qualitative data is descriptive in nature, which does not necessarily lend itself graphical representations. A literature review of close to 800 scientific articles found that only 27% of articles that presented qualitative data used classical graphics to present the information.[2] Hence, presenting qualitative data focuses on the narrative and the story. To do this, we will be going over a few tools that can be used either to reinforce the presentation, or to help illustrate the collected data.

6.2.1 Overview of Presenting Qualitative Data

Presenting qualitative data means focusing on organizing the story into a narrative that makes sense to you and the readers. Hence, focus on the general themes that you have identified during data analysis, and focus on how you got to these conclusions. If possible, triangulate your data: Showing the level of consensus between participants helps support the validity of qualitative data, but avoid presenting elements in a way that overemphasizes quantitative dimensions of your data, especially if you used tools that emphasized exploration. For example, interviewing 10 key opinion leaders will generate valuable information, but you should never use a quantitative argument (i.e., 4 people out of the 10 we interviewed thought that . . .) to formally demonstrate a point. This type of demonstration tends to bring out counter arguments focusing on sample size, rather than embarking on the narrative you have built.

6.2.2 Using Quotes to Reinforce Qualitative Presentations

A presentation based on qualitative data should reflect that the data consists of patterns found in quotes and observations. Hence, it is quite common to use direct quotes to reinforce conclusions during presentations. Quotes can also be used in reports to reinforce quantitative data, to strengthen an element that is being emphasized, or that was discovered during analysis. If you choose to directly quote someone in a presentation or a report, remember to respect confidentially and attribute quotes anonymously (unless express permission to directly quote is given).

In addition, while you can use some briefs quotes to illustrate a point, do not use too many in the same section your document. Avoid over quoting participants as a way of delivering qualitative data: Quotes illustrate the results of the analysis, but do not replace it. Stringing a series of quotations without analysis makes it difficult for the reader to follow (especially if the narrative style shifts from one quotation to the next).

Finally, it is tempting to choose the most remarkable quotes from your interviews to demonstrate a point but remember that these do not necessarily represent the pattern of your data. Choose quotes that represent a concerted point of view and avoid outliers to illustrate a point.

One way to emphasize a quotation is to use a *boxed display*. Boxed displays are text framed within a box. These are used to highlight a specific narrative or quote which is important enough to be put into evidence. It emphasizes something of specific interest and helps separate a concept and a supporting quotation. Figure 6.4 illustrates well how this concept can be used.

6.2.3 Visual Layouts to Display Qualitative Data

The imaginative use of diagrams and schematics can be useful ways to illustrate analytical processes and findings, and to simplify more complex information. While there are many ways to visually display information in qualitative presentations, we will be emphasizing matrixes and flow charts.

> **On the importance of boxed displays ...**
>
> A few years ago, I completed a project with a client in South-East Asia. The project was to identify the feasibility of establishing a small bio industrial complex, based on regional strengths and competition. The report included a number of interviews with companies interested in the region, but these interviews had revealed a number of potential issues. To ensure the issues were noticed by the reader and that they would be addressed, boxed displays (such as this one) were used sparingly throughout the report, emphasizing those latent issues.
>
> *Jean-Francois Denault, Impacts. Ca*

Figure 6.4 Use of a boxed display to share qualitative information (quotes).

6.2.3.1 Qualitative Matrixes

Matrixes are built by crossing two or more dimensions, variables, or concepts of relevance to the topic of interest and to see how to interact. They are used to classify data across topics and demographic data or more complex illustration of results. The advantage of building a qualitative matrix is that when you are building them, you get to understand your data further, giving you an additional level of analysis. Of course, not everyone is visually oriented, but they are an interesting way to display information.

To illustrate, lets refer to a project I did, a survey for a media client active in the Gen Y demographic space. This client was investigating purchasing patterns for health drinks. He had asked participants about the last drink they had purchased, and what had led them to purchase that product over another competing brand or product. Since the objective was purely exploratory, the answers participants gave were open-ended. Once the data was codified, I created a framework with over a dozen different categories of responses. The top five answers had the most responses. To see the difference between purchasing factors (qualitative data) and the sex of the person answering the survey (quantitative data) we built a matrix (see Table 6.2).

The advantage of a matrix is that it has the feel of quantitative data, and is useful to convey messages, especially when you have access to large data samples.

Table 6.2 Example of a Mixed Matrix (Qualitative and Quantitative Data) Showing the Relationship between Purchasing Factors and the Sex of the Respondent

Purchasing Factor	*Answers*	*%*	*Female*	*%*	*Male*	*%*
Taste	689	*68.9%*	336	*67.2%*	352	*70.4%*
Price	366	*36.6%*	182	*36.4%*	183	*36.6%*
Refreshment	278	*27.8%*	137	*27.4%*	140	*28.0%*
Health	271	*27.1%*	156	*31.2%*	114	*22.8%*
Availability	239	*23.9%*	117	*23.4%*	121	*24.2%*

6.2.3.2 Flow Charts

Flow charts are diagram used to document, analyze and illustrate a process or organizational structure using various graphical elements, connected by arrows. They help visualize what is happening, can help illustrate a problem and a possible solution. Also, they can be used to effectively communicate a process to other parties, so they understand it. In presentations, flow charts are useful to illustrate an existing process, for which a product or services resolves a challenge or solves a problem. They can be a useful way to illustrate qualitative data you have gathered from interviews (e.g., how stakeholders usually perform an activity, or how they complete a transaction), and triangulated with other sources of information (e.g., secondary research or observation).

Some conventions around flowcharts include the use of rectangles when illustrating an activity, and a diamond when illustrating a decision. Also, stay consistent throughout your flowchart: A consistent approach eliminates unnecessary distractions and let the reader focus on the essentials. Try to keep shapes the same size as much as possible and limit the use of multiple colors. Finally, try to keep everything in one page: If it gets too crowded, it might be an indication that you have multiple processes or activities intertwined, so try splitting them up into two or more flowcharts, linking them as needed.

When building a flowchart, start by organizing the tasks in the chronological order they happen or following the hierarchical structure. Note when decision points occur, and what the various consequences are. Also, try to take note when there is a feedback loop (an activity that returns to a previous step in your flowchart). Then, do a first draft of your chart: I use a whiteboard as first drafts are often cluttered, and redundant steps may need to be removed. Once you have a stable flowchart, you can use Word or Excel to illustrate it, or a more specialized software or website if you are building something particularly complex.

To illustrate, I refer to a research project I did quite a few years ago. A CEO at an emerging startup was interested in benchmarking personnel assigned to clinical research per product in biotechnology firms in Canada and the USA. As information in this field is usually proprietary (and not available at large), research was conducted through three indirect sources: Review of industry literature, interviews with individuals and review of hiring history for companies of similar size and following a similar business

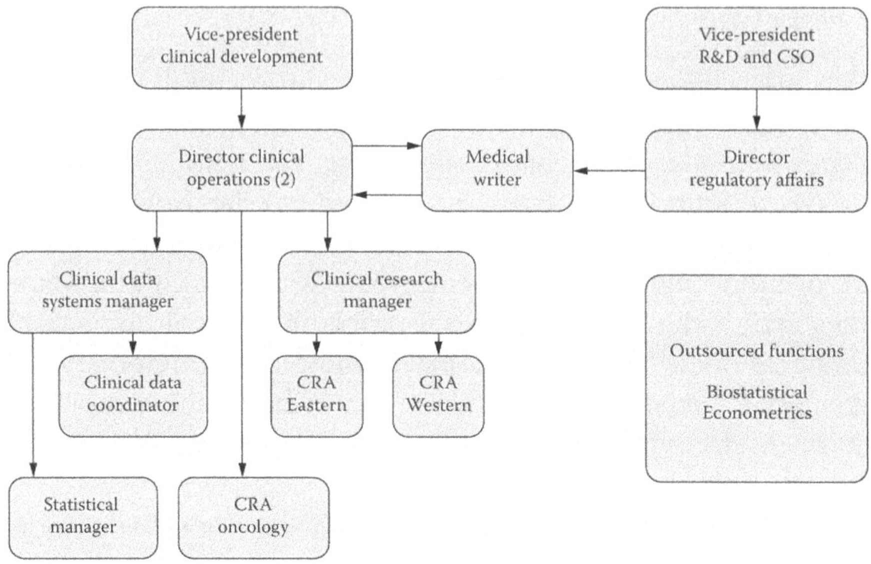

Figure 6.5 Hypothetical clinical development structure for mid-sized biotechnology company.

model. The hypothesis was that by triangulating three independent sources, we would obtain accurate data.

While the results are proprietary, we can share an example of a flowchart (Figure 6.5) that was built from our consolidated data. Already, this gave the CEO some perspective on organizational development and trends.

6.3 Presentation Tools

There are some interesting presentation tools that you can use to present information, especially if you are consolidating different visual layouts, models and information into a coherent story. First, we will take a moment to go over some slideshow basics, followed by an overview of visual storytelling software and infographics.

6.3.1 Slideshow

Slideshows are presentations consisting of information, pictures, videos, and audio shared in a sequential order using an electronic device. They have the advantage of being potentially more attractive and interesting than a traditional paper document and can be exported in standardized file formats (e.g., a PDF) so they can be distributed at large.

While we assume that most people reading this book will be familiar with the basics of doing a slideshow, there are a few fundamentals that you should remember during construction of your deck.

a. *The least text possible:* Do not put every single word of your presentation on the power point slides. Having too much text tends to lead the audience to *read* the information, rather than *listen* to the presentation. As your audience can most likely read faster than you can speak, they will have completely read your slide long before you have shared the information.[3] They will be ready to your next slide, and most likely will not even be listening to you, missing key information or drifting away from your content.
b. *Standardize the fonts and backgrounds:* Use the same fonts (sans serif, if possible) across your entire presentation. Use one font size for your titles, and one font size for your text. A third font sized could be used only if you want to specifically emphasize some key information. Standardized fonts and backgrounds make it easier for the audience to follow without being distracted. An alternative to standardized fonts is using a font matcher (e.g., FontPair (http://fontpair.co/)) to find proven font combinations for both your presentations and your documents.
c. *Use the 6 x 6 matrix:* Synthesize your information as much as possible. Try to keep every slide following the 6 x 6 matrix: A maximum of 6 bullet points per slide and 6 words per bullet point. The objective is to make sure that you have synthesized your information, and you own your content beforehand. This has the added benefit that you will not be able to read your slides mechanically.
d. *Test your presentation onsite:* If possible, arrive early and test your presentation on the computer provided, as well as testing the projector: There is nothing more destabilizing than having technical challenge or a formatting glitch midway thought a presentation. If the content is not too sensitive or confidential, send your presentation ahead and ask the recipient to test the presentation on their hardware to make sure there are no issues.
e. *Color management:* Colors can increase participants' interest, but incorrect use can create distraction or make the presentation ineligible. As a rule of thumb, try to keep cooling colors for the background of your slides (e.g., blue and green), and warmer colors (e.g., orange and

red) for objects in the foreground. Also, be consistent throughout your presentation: Once a color scheme is chosen, keep it for the remainder of the presentation unless there is a specific reason for shifting (i.e., calling attention to a specific element). There are many online tools that can be used to manage colors and chose ones that match best. Personally, I am a big fan of Coolors (https://coolors.co/).

6.3.1.1 Slideshow Software

There are a lot of alternatives slideshow software available on the market. Here are some of the most popular or interesting ones out there:

- *PowerPoint* is by far the most widespread slideshow software. It is well known by many users, is simple to use and can be quickly setup for a presentation (most laptops have PowerPoint installed on them). It also allows you to export slides into PDF format, making it easy to share content in a secure format. It comes with many templates pre-installed, but due to the prevalence of PowerPoint, many audience members will be familiar with most of the standard templates so if you use them, you will not be winning any points for originality.
- *Google Slides* is a powerful alternative to PowerPoint. Like PowerPoint, it lets you integrate most format files into you presentation and integrates well in Google Software Suite. The main strength of Google Slides is its collaborative aspect: It allows multiple users to access and update slides simultaneously. If you are working in a collaborative online environment, Google Slides might be the alternative for you.

6.3.2 Infographics

While classical methods of presenting data are well known and recognized by the community, there is a limited visual appeal to them. In addition, there is very little that is distinctive about them. One of tools to emerge in this space is the use of *Infographics* to present data.

An infographic is a graphical visual display of information and data used so the information can be assimilated quickly by the reader in a pleasant manner. Hence, once you have found or created interesting and reliable data, infographics can be a useful tool to produce engaging visuals and create interest.

The advantage of using Infographics' is that they show overall context in a clear and visually appealing way. Immediately, the viewer can easily understand what the document is about as well as seeing how each data set is related to each other. Infographics are appealing because it is easier for people to process images than it is to process text. Also, they are a modern way of presenting otherwise traditional data and are portable, as you can as easily insert them into a business plan, a webpage, a physical printout for distribution or even post them across social media platforms.

Combing elements of table presentations with graphical elements, infographics allow the user to communicate messages in an attractive way to the audience. But to do so, the user will have to move beyond the traditional software suite. There are several interesting platforms that possess powerful capabilities to build infographics such as Canva Infographic Maker (www.canva.com/create/infographics/).

The limits of infographics are that is that the researcher is limited in the complexity of the information he can share, and that abusive use of infographics can lead to situations akin to "power point poisoning," i.e., the overload of information presented in such a disjointed method that it bores and loses the recipient's interest. As such, try to limit the amount of text on your infographic, privileging images and statistics instead.

One of the other main issues with infographics is data accuracy. If you have one incorrect data point on your infographic, it will quickly impact your overall credibility of the other data and the infographic in general, so make sure your data is completely accurate. Also try to limit the number of colors you use on a single infographic: Too many colors make it difficult for readers to understand the information you are trying to share with them. Finally, while you might have a lot of great information to share, be careful not to overload on content: Excessive information will impact user readability. Making users wait for the page to load, or having to scroll down will reduce the overall effectiveness of your infographic. If you have too much content, consider splitting the information into two separate documents, rather than trying to cram everything in a single document.

As an example, I have included an infographic I prepared for Ancon Medical many years ago, a client active in the early cancer detection space. This is based on the information found in his business plan as well as different market research reports I prepared for them. Consequently, I was able to build an infographic (Figure 6.6) that conveys information on the importance of early cancer detection as well as showcasing some of the key relevant market data we had found that supported the company's vision.

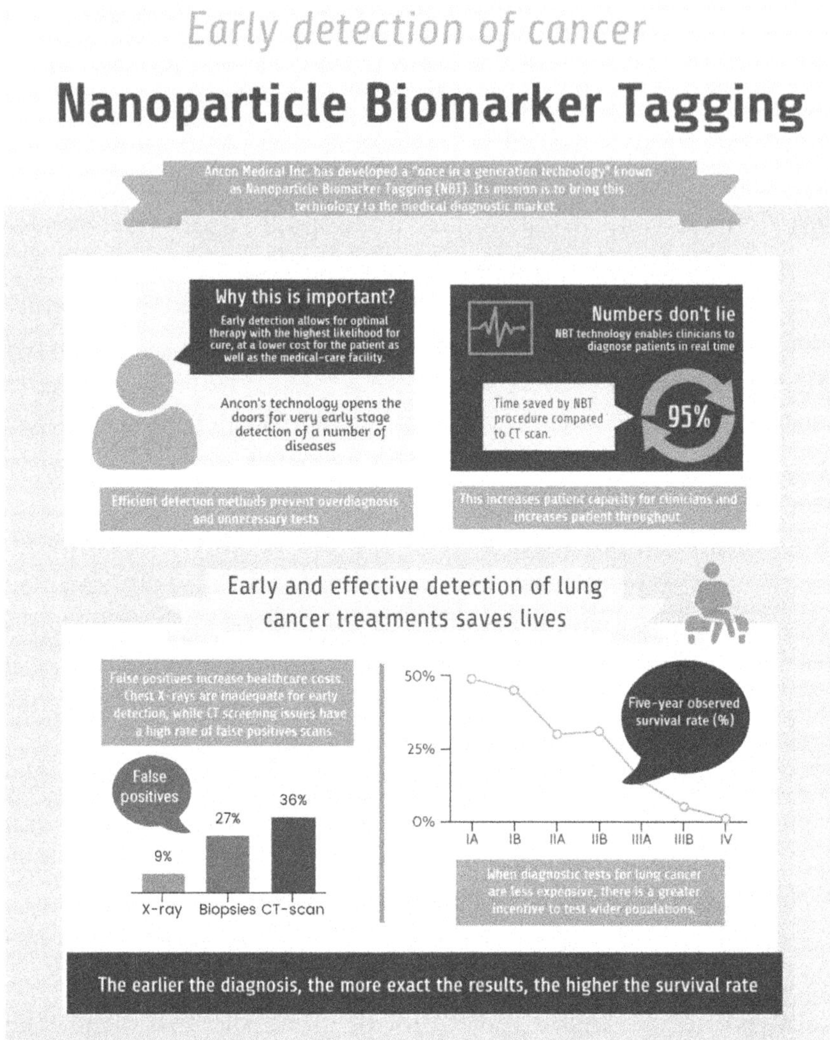

Figure 6.6 Example of an infographic for a company in the early cancer diagnostic space.

6.4 Slideshow Framework: The 10/20/30 Rule

The 10/20/30 rule of presentation was proposed by Guy Kawasaki, a Silicon Valley venture capitalist specialized in marketing. Having seen numerous pitches and presentations, he noticed that entrepreneurs were concentrating on unimportant content and losing focus. Since then, he has been advocating the 10/20/30 Framework. The objective is for the presenter to synthesize his corporate presentation down the essentials: "A PowerPoint presentation should have ***ten slides,*** last no more than ***20 minutes***, and contain ***no font smaller than thirty points.***"[4] A complete description of his approach is available on his blog. The ten slides that he deems essential are:

1. The Problem: What is the problem you have identified?
2. Your solution: How are you fixing the problem?
3. Business model: What is your organization business model?
4. Underlying magic/technology: One slide dedicated to your technological advantage?
5. Marketing and sales: What are your market, and how do you expect to generate sales?
6. Competition: What is the current competition (technologies, companies)?
7. Team: Who are the key members, and their strengths?
8. Projections and milestones: What are the key milestones to your success?
9. Status and timeline: What are the current status of your technology, and your timeline?
10. Summary and call to action

Afterwards, the onus of the effort is on the presenter to fit all his information in a 20-minute presentation. As mentioned by M. Kawasaki, "In a perfect world, you give your pitch in twenty-minute, and you have forty minutes left for discussion."

If you do feel you have more information, put them in your annex. Back-up slides are a great way to feel confident that you can address specific points without drowning your listener in too much information. Finally, he recommends using 30-point fonts, as he believes that too much text means that you will be reading your text rather than presenting your presentation. It forces the speaker to know his content, and to synthesize on the essentials.

6.5 A Few Tips on Storytelling

6.5.1 The Three What's of Storytelling

Presenting your company is about taking your audience through the three **what's**:

1. What happened? (The problem),
2. So what? (Why is this important)
3. What now? (How are we solving this).

In life, stories create recognizable patterns. Be it a movie we are watching, a book we are reading, or even a presentation we are listening to, our brains look for patterns. They look for continuity. If you can write a compelling story, you will find your audience that much more invested in your pitch. Long after content itself is forgotten, the stories you tell stand a chance of being remembered, and without a story, it is hard for the audience to navigate or understand what this means to them.

For example, contrast the two following pitches:

- *Pitch #1 (Science-driven):* A Phase II clinical trial ($n = 127$) demonstrated that the combination of our immunotherapy agent with standard chemotherapy resulted in a 34% improvement in progression-free survival ($p < 0.05$) compared to chemotherapy alone in patients with advanced non-small cell lung cancer.
- *Pitch #2 (Story-driven):* When Sarah's oncologist told her the tumor had grown despite 6 months of chemotherapy, she thought she had run out of options. But a new approach, using our new compound as an ally alongside the chemo bought her something she had almost given up on, time. Eighteen months later, her scans remain stable. She was one of 127 patients whose results suggest this combination might be the next step in lung cancer treatment.

The first presentation provides information, while the second one shares a story, one that will stay will listeners long after the presentation.

6.5.2 The Law of Seven

George Miller's law of human cognition is a useful rule of thumb when presenting to an audience for the first time. He proposed that we can only hold approximately seven pieces of information in working memory at once. Beyond that, we start selecting what to remember and what to discard. While the exact number has been debated (some argue it is closer to five), the principle holds: Human cognitive bandwidth is limited.

So rather than overwhelming your audience with data, analyze and select the most important points you want them to remember. For complex stories, group related data into a few manageable themes and explain why these matter most.

This is why repetition works. If you want people to remember something beyond the presentation, build your talk around a handful of key pillars and circle back to them at the end of your presentation.

6.5.3 Simplifying Your Technology Story

I recall attending a client's pitch a few years back. He had prepared his slide deck and felt quite confident about his presentation. He had invited me to attend to a pitch he was making as to get my opinion on potential investors and to give him feedback on his presentation skills. He was quite proud of his technology. So proud, in fact, that his presentation focused almost exclusively on the technical innovation and prowess of this technology, with very little focus on the market, commercialization plan, or business model. When he finished, one of the investors candidly asked, "What are you selling, exactly?" After that presentation, I must admit I was as confused as the investor.

On multiple occasions, I have been asked by clients to review and comment their "pitch deck." One thing I constantly point out is that these slide decks are far too technically focused, with many slides dedicated to the science but addressing the audience from such a specific angle that it is unintelligible to the average person. While your technology is undeniably a key element of a company's unique selling proposition, if the audience is unable to understand the technology, it will quickly become "white noise."

While some individuals possess exceptional research skills, creating paradigm shifting technologies and redefining markets, they must remember that not every person in the audience might possesses the same level of skill. Hence, it is possible that the message never reaches the recipient as he is just unable to understand it. It is important to take note of the audience's knowledge level and customize the message to match their level of understanding. You might have to explain what you feel are basics terms and concepts, so be ready to describe them. If possible, develop stories, examples, and parallels to simplify complex technologies.

Overall, in the initial meetings, emphasize must be given to the market opportunity that the technology provides, and how the company intends to approach the market to commercialize the technology. Later, if there is interest, more time will be made available during due diligence to demonstrate scientific and technical validity of the technology.

Notes

1 Denault, J.-F. 2014. How much do environmental attributes influence purchasing patterns? *Marketing Magazine*. https://www.marketingmag.com.au/hubs-c/how-much-do-environmental-attributes-influence-purchasing-patterns/#.U4MYzfldXL8 (Accessed January 26, 2026).

2 Verdinelli, S., & Scagnoli, N. 2013. Data display in qualitative research. *International Journal of Qualitative Methods*, 12, 359–381.
3 If you speak faster than attendees can read your presentation, you possibly have another presentation issue to address. Just saying.
4 More information on the 10/20/30 rule is available at http://guykawasaki.com/the_102030_rule/ (Accessed January 26, 2026).

Chapter 7

Final Words

While you can have virtually anything you want, you cannot have everything you want.

—Ray Dalio

Many years ago, as I was completing my graduate studies in communication, we were working on a case study related to crisis management. Hidden inside the case study was a simple throwaway line from one of the participants, who mentioned in passing that **"Market Research creates our reality."** To this day, I still think of this statement, as I believe that it candidly describes the importance that market research has on how the startups grow, pivot, and ultimately develop.

Business leaders and decision-makers require high-quality information if they want to make accurate decisions. As we mentioned multiple times in this book, market research is often performed when the cost of doing the research outweighs the cost of not doing it, and the dangers of making uninformed decisions run the risk of being costlier than doing nothing. Decision-makers need that market data as part of their decision process.

From investors to future hires, everybody needs to know what the market is like. Investors need to know because they want to be sure that they are investing in the right place. Individuals need to know because they want to be sure that they are investing their most precious resources—their time—in the right place, not in an organization that will go belly-up in a few years because it was building a product nobody needed. Everybody in the business ecosystem is seeking for the same thing, the answer to the question: "Does the market exist?"

DOI: 10.4324/9781003731726-7

And it is market research that brings this world to life.

Starting from a hypothesis that there is a market, somewhere out there, the market research you will be conducting using the tools gathered from this book will help to understand that story, find out just how many people need your product, and what exactly they need. From laboratory to market, the journey to learning your market can be iterative, and must be ongoing, continually measuring your competitors' innovation through competitive intelligence activities to reaching out to your customers through voice of customer activities.

So where does that leave us?

As technology grows at a furious pace, so does our little corner of the market research world. Many of the tools and techniques that were common in the first edition, are now classified as "traditional,": Still useful, but slowly being replaced by new technologies that revolutionize many of the approaches that were all too common a few years ago.

Furthermore, the emergence of artificial intelligence has considerably impacted our approach and ability to conduct market research. There is a great temptation to give these tools more space and responsibilities in data collection, and to trust the results blindly. After all, it does everything faster than we ever could. But there is an underlying reality; all AI outputs are hallucinations: Some just happen to be better and closer to reality than others. And while AI may have near-perfect accuracy in some domains (like basic arithmetic or well-documented facts) for others (such as recent events or consumer behaviors) they do not have the common-sense abilities humans have, making generated data unreliable if you are unable to double-check the statements yourself. The need for your own collected data (be it interviews, surveys, focus groups) remains crucial to contextualize and verify data. In the end, humans and AI are far more likely to work together because that produces the best results, rather than having a perspective where AI is replacing humans.

While attempting to be contemporary, I have reviewed and suggested several software programs and platforms to assist in your market research, analysis, and classification. As you are reading this, some of the tools might already be defunct, replaced by newer and faster alternatives or simply replaced by a generic AI tool. Case in point, during the preparation of this second edition, I found myself spending almost as much time fact-checking and verifying that the tools I suggested still existed and worked in the way I had initially wrote than I did generating new updated content. Such is the curse of living in such a fast-paced world. Today's tool is tomorrow's artifact.

I still believe that tomorrow's market research effort will most likely depend more on passive secondary research tools, as these become more efficient and include learning algorithms to generate better searches and pre-analyze data. In the first edition, I believed that as technology improved, and people's knowledge of information technologies increased, market research would most likely shift to include more web-based online components, such as online focus groups and online interviews, slowly replacing traditional in-person focus groups and telephone interviews. This is something that has come to pass, as most interviews today are conducted using tools like Zoom and Microsoft Teams for convenience as they integrate recordings as well as the ability to generate both transcripts and summaries. As for focus groups, this transition has been somewhat slower as online tools they still do not replicate the richness of data that live interactions generate. Yet.

The reliability of social media as a research source has also quickly deteriorated since the first edition of this book. A few years ago, people posted their genuine thoughts and experiences, making platforms valuable for understanding consumer sentiment and behavior. Today, feeds are increasingly filled with AI-generated posts and blogs created not to share insight, but to hit publication metrics. LinkedIn is the great example of this, often flooded with AI-generated "inspirational" posts, fake "lessons learned" stories, and engagement bait. Twitter/X now has bot farms dedicated to generating content. Instagram now has AI-generated influencers telling us what product they love. "Authentic content" has genuinely collapsed in the last 18–24 months, making "traditional" social listening a challenge to collect.

When AI is utilized lazily—for example, write me a post about leadership—the output is generic, keyword-filled content with no authentic experience behind it. Even thoughtfully prompted AI (e.g., synthesize my key learnings into two points, emphasizing XYZ) still produces content optimized for engagement rather than sharing true experiences. This flood of artificial content means social media data now requires far more scrutiny. Researchers must develop new filtering skills to distinguish genuine human insight from AI-generated noise, or risk building strategies on artificial sentiment.

And while this book emphasizes your own data collection, I must take a moment to remind you of the limitations of self-collected data when raising funds. This data can be perceived as biased, and since methodology is often unstandardized, there is a limited value giving to it. Independent evaluation might be needed at some point, either through your own initiative or by the investor or stakeholder evaluating your technology. As such, the main objective of collecting your own data should be to assist you in your own decision-making.

Nonetheless, however technologies change, market research will continue to play a role in how organizations shape their interaction with their clients, their stakeholders, and their competitors. The ability to reach out and talk with your current and future clients will let you build products that fit not only a perceived market but also one you have discovered, verified, and validated.

It is my hope that this book gives you the tools you need to understand and share your story.

Appendix A The Market Research Checklist

Sometimes, you want your research to go beyond the standard question of: "What is my market?" To assist during the brainstorming and planning phase, here is a short list of question that can be useful to orient your research efforts. Questions break down into three distinct categories: Your Corporation (Market and Product), Your Consumers, and Your Competitors.

Your corporation

- What is the size of the market?
- Is the market growing, steady, or declining? At what rate?
- Is it an established market or an emerging market?
- Are there any changes in the environment that may impact your product, such as government legislation, regulatory changes, taxation, or more?
- Is your product a disruptive innovation or an incremental innovation?
- What is the best way to promote the product?
- What should your commercialization strategy be?
- What should your pricing strategy be?

Your Customers

- Who is the purchaser of the product or service? Who is the decision-maker?
- What is their purchasing process? How can you reach them?
- Who uses the product or service?
- Is the purchaser different then the end-user? How do these two stakeholders interact?
- What is the product or service replacing?
- What is the user presently using to fill their current need?
- What do potential customers or end-users think of the product or service?
- What is the profile of your customers (location, age, gender, income level, etc.)?
- What are their needs? What need does the product specifically target?
- What are the customer service and retention strategies that are needed?

Your Competitors

- What are the products that compete directly with your product? Indirectly?
- Who are the main competitors and what share of the market do they have?
  - What are their strengths and weaknesses?
  - Where do they have a competitive advantage?
  - What is their established pricing strategy?
  - What is their branding strategy or Unique Selling Proposition (USP)?
- How can you differentiate your product from competing products?
- How is the market shaped?
  - What should your market segment be?
  - Are you positioning yourself correctly?
- How do potential customers compare you with your competitors?

Index

Note: Page numbers in *italics* indicate a figure and page numbers in **bold** indicate a table on the corresponding page.

L

M

N

O

P

For Product Safety Concerns and Information please contact our EU representative GPSR@taylorandfrancis.com
Taylor & Francis Verlag GmbH, Kaufingerstraße 24, 80331 München, Germany

www.ingramcontent.com/pod-product-compliance
Ingram Content Group UK Ltd.
Pitfield, Milton Keynes, MK11 3LW, UK
UKHW050058200926
475702UK00019B/2904
* 9 7 8 1 0 4 1 2 2 3 2 2 1 *